Botanical Research and Practices

Botanical Research and Practices

Oregano – The genus Origanum (Lamiaceae): Taxonomy, Cultivation, Chemistry, and Uses
Tuncay Dirmenci (Editor)
2021. ISBN: 978-1-68507-315-2 (Hardcover)
2021. ISBN: 978-1-68507-387-9 (eBook)

Punica granatum: Cultivation, Properties and Health Benefits
Rupesh K. Gautam, PhD (Editor)
Smriti Parashar (Editor)
2021. ISBN: 978-1-53619-767-9 (Hardcover)
2021. ISBN: 978-1-53619-821-8 (eBook)

Elementary Botany
George Francis Atkinson (Editor)
2021. ISBN: 978-1-53619-448-7 (Hardcover)
2021. ISBN: 978-1-53619-518-7 (eBook)

Brassica juncea: Production, Cultivation and Uses
Dhriti Kapoor (Editor)
Vandana Gautam (Editor)
2021. ISBN: 978-1-53619-241-4 (Hardcover)
2021. ISBN: 978-1-53619-289-6 (eBook)

Passiflora: Genetic, Grafting and Biotechnology Approaches
Alejandro Hurtado Salazar
John Ocampo Pérez
Nelson Ceballos-Aguirre
Dora Janeth García Jaramillo
Walter Ricardo Lopez
2021. ISBN: 978-1-53619-108-0 (Hardcover)
2021. ISBN: 978-1-53619-255-1 (eBook)

More information about this series can be found at https://novapublishers.com/product-category/series/botanical-research-and-practices/

Mamta Pujari, Bharat Kapoor,
Neeta Pande and Anju Joshi
Editors

Coriandrum sativum

Origin, Uses and Nutrition

DOI: https://doi.org/10.52305/TXFR4149

NOTICE TO THE READER

Library of Congress Cataloging-in-Publication Data

ISBN: 978-1-68507-842-3

Published by Nova Science Publishers, Inc. † New York

Contents

Preface

Coriandrum sativum is one of the intriguing annual herbaceous medicinal plants which is raised in the Mediterranean area. However, it is copiously grown in Asia, Europe, and North Africa as a medicinal and culinary herb, which belongs to the Apiaceae family. It has diploid chromosome numbers (2n=22). Its seeds are commercially utilized as a condiment in many areas of the world or as a principal element of Indian curry food. Numerous ingredients, such as fish and steak and bakery and confectionery items, are flavored with seeds. The fresh leaves usually referred to as cilantro or parsley, are widely utilized as a food for flavoring or disguise the irritating smells of specific foods in eastern cooking and Indian cuisine. It's interesting to note that all portions of this plant have been employed as herbal medicines in western medicine to cure various ailments. Coriander seeds are often used to diagnose a range of digestive issues, including dysentery, indigestion, and nausea, while its leaves energize hunger and aid. It was discovered that Coriander has many drug-related effects such as anti-proliferative, anti-hyperglycemic, hypotensive, anti-hyperlipidemia, anti-perpetuity, digestive stimulants, anti-fertility, and hypotensive.

Coriander is produced in various agroecological conditions, mainly at smaller production areas. The most important producers of coriander in the world are: India, Mexico, Iran, Morocco, Canada, Romania, Russia and Ukraine. Coriander is rich in bioactive compounds which are present in the entire plant. These constituents are usually isolated from ripe fruits or the above ground parts of the plant with inflorescences. Coriander is very rich in nutrition values and scientific reports suggest that whole-body parts of the plant are composed of different vitamins, minerals and have good moisture content. It is observed that vitamins like riboflavin, niacin, ascorbic acid (vitamin C), retinol (Vitamin A) are present in the leaves of coriander. Fresh leaves of coriander are reported to have minerals like calcium (0.14%), phosphorus (0.06%), and iron (0.01%).

The whole seeds of coriander are an integral part of spice blends or curry powders along with other spices such as black pepper, chillies, cumin, fenugreek, fennel, onion and garlic etc. Seeds when grounded coarsely have a crunchy texture and they form the bulkiest component of curry powders. This seed powder is used in brining, pickling and seasonings. Out of all the fatty acids in fruits, petroselinic acid (cis-6-octadecenoic acid) is the principal component imparting unique physicochemical properties followed by linoleic acid, myristic acid, palmitic acid, stearic acid, oleic acid and vaccenic acid. Oil cakes left after oil extraction are quite rich in carbohydrates like cellulose, proteins, fats, some nitrogen free extracts and ash which could be employed as animal feed.

Chapter 1

Origin, Status and Cultivation of *Coriandrum sativum* L.

Tunisha Verma[1,*], Savita Bhardwaj[1,*], Dhriti Sharma[1], Vandana Gautam[2] and Dhriti Kapoor[1,†]

[1]Department of Botany, School of Bioengineering and Biosciences, Lovely Professional University, Phagwara (Punjab) India

[2]College of Horticulture and Forestry, Dr. Y. S. Parmar University of Horticulture and Forestry, Nauni, Solan (HP), Neri Campus (Himachal Pradesh), India

Abstract

Coriandrum sativum L. belongs to *Umbelliferae* family and is an annual herb and medicinal plant. All the plant organs of coriander are edible; however, their flavors and application are quite interesting. The spice is light and has unique aroma and taste. In the folk medicine systems of several cultures, the root, stem, leaves, and fruits of this plant are employed as flavoring agents or traditional remedies for the treatment of various illnesses. It is extensively cultivated worldwide since prehistoric times and is commercially cultivated presently in Morocco, Romania, France, Spain, Italy, the Netherlands, Myanmar, Pakistan, Turkey, Mexico, Argentina, India and Western Australia and to some extent in the countries like UK and USA. This plant is rich source of numerous bioactive constituents like essential oils, fatty acids and lipids, which are extracted from the aerial sections of the plant. A wide range of pharmacological activities has been found in different parts of this herb

[*] Equal contribution.

[†] Corresponding Author's Email: dhriti405@gmail.com.

In: *Coriandrum sativum*: Origin, Uses and Nutrition
Editors: Mamta Pujari, Bharat Kapoor, Neeta Pande et al.
ISBN: 978-1-68507-842-3

due to the presence of numerous bioactive constituents, including anti-microbial, anti-oxidant, anti-diabetic, anxiolytic, anti-epileptic, anti-depressant, anti-mutagenic, anti-inflammatory, anti-dyslipidemic, anti-hypertensive, neuro-protective diuretic, and hormone balancing effect that promotes health. The objective of this chapter is to emphasize coriander's origin, history, processing, and functional properties.

Keywords: herb, medicine, traditional, bioactive constituents, nutritional

1. Introduction

Spices play a significant role in our national economy and are recognized as one of the most effective therapeutic foods due to their health benefits as traditional medicine since ages (Bakkali et al., 2008). The essential oils and extracts of spices have been utilized in food preservation and pharmaceuticals (Christaki et al., 2012). The ability of spices to impart biological activity is slowly becoming a topic of attention in the field of medical science (Properzi et al., 2012).

Coriandrum sativum L. is a significant plant that originated in the Mediterranean region and is widely cultivated as medicinal and culinary plant across the world especially in North America, Africa, Asia and Europe (Laribi et al., 2015). Apart from this, it is extensively cultivated in different environmental conditions as Chinese parsley or cilantro. It is an annual herbaceous plant belonging to family *Apiaceae* with 50 to 100 cm height. It can be easily be distinguished by its broad green lanceolate-shaped leaves, prominent taproots, and by white, pale or pink umbellate asymmetrical flowers (Mandal and Mandal, 2015). All plant organs of coriander are utilized to make a popular spice with a pleasant lemony flavour, including the leaves, stems, seeds, and roots (Marouti et al., 2010). Coriander occupies prime position in eastern culinary system due to its unique aroma and flavor and has a long history as a source of aroma compounds and essential oils with diverse biological activities. Leaves and seeds of coriander have unique aroma and are used in preparation of different cuisines (Laribi et al., 2015) The yield of *C. sativum* L. essential oil and its chemical composition undergoes changes during ontogenesis, which affects the aroma of the plant (Singletary, 2016).

Coriander's nutritional worth should not be underestimated, since it includes high amounts of vitamin C, vitamin A, vitamin B12, folate and polyphenols, in addition to its smell and flavour (Daly et al., 2010; Puthusseri et al., 2012, 2013, Beyzi et al., 2017). Whereas, coriander fruit is rich in numerous fatty acids such as petroselinic acid, linoleic acid, oleic acid, palmitic acid, and stearic acid.

In terms of health benefits, coriander has been utilized as a folk remedy since the Greek era, according to recent historical documents (Grieve, 1971). It is used as an effective carminative, aromatic stimulant, appetiser, diuretic, antibilious, refrigerant, and aphrodisiac, as well as a digestive stimulant which stimulates the stomach and intestines. The seeds of coriander are used to cure wide range of issues related to indigestion, nausea, dysentery, chest pain, coughs, rash, fever diarrhoea and post- partum complications (Yorek, & Dogan, 2009). Constituents of coriander are known for their health promoting, curative and preventive properties like hypoglycemic, hypolipidemic, antihypertensive, antibacterial, antihelminthic, and antimutagenic properties (Laribi et al., 2015). The entire coriander plant is used in traditional Ayurvedic system (Yadav et al., 2015). Since now, essential oil and biologically active components present in coriander have been used in perfumes and cosmetics (Draelos, 2013).

Given the potential use of medical herbs as functional ingredients, it's exciting to put together a comprehensive review that covers the nutritional, medicinal, and bioactive aspects of coriander. As a result, this review examines the medicinal, phytochemistry, and pharmacological properties of several components of this unique herb in order to investigate its potential usage in the functional food and nutraceutical industries.

2. Origin, History and Distribution

Hippocrates, who lived around 5000 BC, used coriander as one of the first herbs in Greek medicine. It is mentioned as far back as 5000 BC in Sanskrit literature, and as early as 1550 BC in Greek Eber Papyrus (Prakash 1990). In Sanskrit literature, coriander is known as "kusuthumbari" or "dhanyaka." In Israel's Nahal Hernel cave, mericarps were discovered in a Neolithic mix, which may be the oldest archaeological findings of coriander. In the tomb of Tutankhamun in Egypt, one-half of coriander seeds were discovered. Because of its biological properties, the herb was named as "spice of happiness" in Egypt (Zohary, 1993). Coriander is mentioned in the Bible in Exodus 16:31, and was disseminated throughout Europe by the Romans. It was cultivated in Greece from at least the second millennium BC. It is said to be one of the first herb to be cultivated in North America in Massachusetts around 1670 from where it spread to Latin America (Germer, 1989).

Coriander is one of the first spice used by mankind and has been cultivated since antiquity (Small, 1997). Seeds of coriander were first discovered by Egyptians around 3000 years ago (Holland et al., 1991). This herb is found wild in Egypt and Sudan and is native to Middle east and Mediterranean region. The cultivation of coriander is clearly mentioned in the bible and dates back around 5000 years ago. In Roman cultures it was used as a spice to preserve meat and flavor breads and they also introduced this plant to Britain, where it was widely used as cookery until Renaissance, later on it was replaced by new exotic spices available. Coriander is also used an aromatic stimulant due to its medicinal properties by early physicians. In India this plant has been known since Vedic period, until Mughals arrived in India who extensively used seeds of coriander for the preparation of Mughlai dishes.

In tropical region, two varieties of coriander are extensively cultivated including *C. sativum* L. and *C. tordylium.* Coriander origin is unknown however most experts believe it came from the Near East and spread to parts of central Asia, Abyssinia and Mediterranean countries (Vaidya et al., 2000; Meena et al., 2014; Laribi et al., 2015). Coriander is now grown in Argentina, Brazil, India, Italy, Mexico, Morocco, Netherlands, Peru, Poland, Romania, Somalia, Spain and north Africa (Khan et al., 2014). Commercially coriander is produced by Poland, Romania, Czech Republic, Guatemala, Mexico and Argentina. Coriander cultivation throughout multiple continents demonstrates the crop's tolerance to a variety of agro-ecological conditions as well as its economic importance in a variety of societies around the world. India is the world's leading exporter of coriander with area of 5.25-105 hec and yield of 3.10 -105 ton (Peter, 2004). Ukraine and India control the majority of the coriander oil market. This herb is widely known for its significant economic value because it is utilized as flavoring agents in food and cosmetics (Opkyde, 1973).

3. Botanical Description of Coriander

Coriandrum sativum L., belongs to the family *Apiaceae* (*Umbelliferae*), is majorly grown throughout the year by its seeds (Mhemdi et al., 2011). *C. sativum* L. and its wild relative *C. tordylium* are the only two known genera of coriander. Depending upon the environmental conditions, plant height can range between 20 and 140 cm. Young leaves of coriander are used to flavor sauces and curries in Indian, Chinese and Mexican cuisines, with ternate-pinnate leaves, which are not yet separated into thin straight sections. Erect, thin, sympodial, monochasial and branched stem occurs in coriander with numerous side twigs at the bottom node. Inflorescence occurs at the end of each branch (Sahib et al., 2013).

Leaves of coriander are oval, slightly lobed, consists of several branches and sub-branches (Khan et al., 2014), and aerial leaves are elongated while new leaves are oval in shape (Kassahun 2018). Portion of the upper leaves are straight and more divided. White colored flowers are found in coriander with somewhat brinjal like shades (Mengesha and Getinetalemaw, 2010). Globular fruit consisting of two pericarps, with a diameter up to 6 mm and with mericarps which is generally combined by their borders to form a cremocarp. This cremocarp is brownish-yellow or brown in color, glabrous, and about 2-4 mm in diameter (Bairwa et al., 2017). Flowers of this herb are organized in small umbels, which are asymmetrical with petals that directs away from the middle of the umbel being longer (5–6 mm) as compared to those who directs towards the middle of the umbel (only 1–3 mm long). Dry fruit of the coriander is called as coriander seeds and lemony citrus flavor exists in coriander seeds, generally designated as warm, nutty and spicy (Shukla and Gupta, 2009).

Spindle shaped roots are found in coriander. Convex longitudinal vittae of the plant contain the essential oil of the seeds which is responsible for the plant's characteristic 'bug' smell. Coriander plant name is in fact taken from the Greek word, 'Korion' which means bug. With the increase in the maturity level, odor profile of various plant parts changes significantly. Smell of ripe fruits is very different from the green leaves and unripe seeds (Ceska et al., 1988; Diederichsen, 1996; Small, 1997).

More than 10 g weight was observed for 1000 fruits of *C. sativum* L. var. *sativum* with fruit diameter >3 mm, whereas less than 10 g was found in *C. sativum* L. var. *microcarpum* DC, for the same weight of 1000 fruits with fruit diameter of 3 mm (Ravi et al., 2007). Steam distillation is used for the extraction of coriander oil which is essential oil, whereas solvent extraction is used for the coriander oleoresin, a brown–yellow to green liquid, from the dried fruits of *C. sativum* L. (Anon, 2009).

4. Cultivation of Coriander

4.1. Climate and Soil

Cool and comparatively dry frost-free climate is the requirement of coriander, for the production of good-quality and high seed yields mainly at seed production and flowering phase of plant and it is a tropical crop. Seed production dramatically decreases when frost is followed by the flowering phase (Sharma and Sharma, 2004; Abdella et al., 2018). Increased sterility and declined yield are found during anthesis and seed formation stage due to the presence of high temperature and high wind speed (Sharma and Sharma, 2012). Number of aphids and diseases increases during the flowering time due to cloudy environment. Coriander can be cultivated on almost all types of soil, but soil should be rich in organic matter and well drained. Moreover, its cultivation is not suitable for saline, alkaline and sandy soils. Loamy to moderately heavy soils are suitable for its cultivation as an irrigated crop while conserved moisture on black cotton or heavy soil with greater moisture retention potential is suitable for its sowing as a rain-fed crop (Verma et al., 2011).

4.2. Field Preparation and Sowing

4.2.1. Field Preparation

Two or three plowings are generally required to brought coriander field to a fine tilth, to which soil turning plow should be the first one. Soil may be irrigated lightly before plowing if moisture level of soil is low. Field preparation should be performed when optimum moisture level occurs in the soil for a rain-fed crop, subsequent to the preceding rains. For this, soil should be carefully planked to evaluate the moisture losses until sowing.

4.2.2. Sowing Time

Optimum temperature of 20-25^{0}C is suitable for the germination and early development of coriander. Crop is majorly sown from the end of October month to the middle of the November in northern India (Meena et al., 2016). If the crop is sown earlier, then high temperature conditions harmfully impact seed germination and early crop growth. After the cultivation of coriander is done in the winter season, a second crop which is sown between mid of May to mid of June, is also taken during the summer, in southern India (Sharma and Sharma, 2012). However, plant growth decreases and occurrence of diseases and pests increases due to the delay in sowing.

4.2.3. Seed Rate and Sowing Method

Depending upon the seed size, seed rate of 12–15 kg per hectare is sufficient to achieve optimal crop density in irrigated fields (Moniruzzaman et al., 2013).

But for bold-seeded type, slightly higher seed rate is sufficient. 25–30 kg per hectare seed rate is suggested for sowing in rain-fed conditions. Seeds are treated with carbendazim at a rate of 1 g kg^{-1} of seed or with other 2 g mercurial fungicide per kg of seeds after crushing or dividing seeds into two halves.

Scattering or broadcasting methods are used for the sowing of the treated and crushed seeds and also by sowing in lines apart by 30 cm in shallow rows behind the plow. Intercropping technique is facilitated by line sowing in the standing crop (Jan et al., 2011; Katiyar et al., 2014; Sharma et al., 2016). Row spacing of 40 cm is normally suggested for sowing in heavy soils or under greater soil fertility environments. An ideal distance of 10 cm is recommended between plants in rows. Both methods should be carefully performed to ensure uniform covering of seeds with soil not deeper than 4 cm.

4.3. Plant Management

4.3.1. Manure and Fertilizers

Around 10–20 tonnes per hectare of farmyard manure (FYM) or compost should be practiced on coriander crop during field preparation (Ahmad et al., 2017). In addition to this, during sowing time, 20–30, 30 and 20 kg of nitrogen, phosphate and potash, respectively per hectare should be applied in the form of fertilizers (Sharma and Sharma, 2012). Furthermore, an extra quantity of 40 kg nitrogen per hectare should be added in 2 equal parts with irrigation, first one is at 30 days and second is at 75 days after sowing in irrigated conditions (Carruabba, 2009). However, depending upon the availability of soil nutrients, quantities of fertilizers can vary, through their evaluation with soil testing.

4.3.2. Hoeing and Weeding

Slow growth rate is found in coriander initially. After 30 days of sowing, hoeing and weeding should be performed, as well as point thinning may also be done to remove excessive plants (Kassahun, 2020). Depending upon the appearance or re-growth of weeds, 2nd hoeing and weeding can be performed between 50 and 60 days after sowing. If manual removal of weeds is not possible then weeds can be controlled by the use of chemicals such as by the application of fluchloralin herbicide at the rate of 0.75 kg per hectare before planting, or through a pre-emergent application either of oxyfluorfen or of pendimithalin at the rate of 0.15 and 1.0 kg/ha respectively.

4.3.3. Irrigation

Around 4 to 5 irrigations are needed after germination of seeds depending upon the particular variety used, moisture retention potential of soil and environmental conditions (Zamora et al., 2019). The first, second, third, fourth and fifth irrigation should be given at 30–35, 60–70, 80–90, 100–105 and 110–150 days after sowing, respectively. However, in addition to these scheduled irrigations, sometimes a light irrigation may also be needed between 5 and 8 days after sowing to increase germination process (Sharangi and Roychowdhury, 2014).

4.3.4. Harvesting

It is difficult to know the ideal time for harvesting of coriander because flowering and fruit maturity cannot be exactly assumed. Quality of the spice and essential oil is determined by the harvesting of the fruit at the correct stage of maturity (Sharma and Sharma, 2012). Although content of essential oil is maximum in immature fruits, but consumers are disagreed from their aroma. Typical sweet and spicy aroma of coriander only develop after the fruit has reached maturity stage (Shashidhar et al., 2013). There is a risk of shattering of fruits on the primary umbels if there is delay in the harvesting to permit secondary umbels to gain maturity. Seed shatter is most likely to occur in windy conditions, and even complete loss of umbel can occur. Proper harvesting time must be arbitrated very carefully with the progressive fruit ripening. Generally, harvesting should be carried out when half of the total seeds have transformed their color from green to gray–green or yellow to get a good seed quality with highest yield (Balasubramanian et al., 2016). Coriander plants should be cut either early in the morning or late in the evening in order to minimize the damage. Coriander seeds should be dried in order to reduce moisture content to around 9%, which is more than 20% when harvested (Smallfield et al., 2001). Harvesting by using a harvester–thresher should be carried out early in the morning when the seed moisture content is less than 15%. This is due to the fact that after the drying of dew, fruits fall off easily.

To maintain the proper seed color and quality, then the harvested plants should be dehydrated in the shade. If this is not possible, then the plants should be set aside in bundles upside down in order to protect them from direct sunlight, because sunlight impact the color of the seeds (Balasubramanian et al., 2016). Artificial drying of coriander at 80–90°C temperature, is also performed in some countries, for example in Russia. Volatile oils of plant lost when temperatures exceed over 100°C.

Winnowing or light beating with sticks are generally used for the separation of seeds after the drying of harvested plants. Dried bundles should be threshed on a clean floor or on tarpaulin and they are packaged in jute fabric bags after drying and separation.

5. Active Constituents of Coriander

Coriander's nutritional value is divided into two categories: major bioactive constituents and minor bioactive constituents. Essential oils and fatty acids are the major bioactive constituents. Whereas the minor bioactive constituents consist of Monoterpene hydrocarbons viz α- pinene, β-pinene, limonene, γ-terpinene, ρ-lymene, borneol, citron, Xmphoe, Geraniol and Geranylacetate, Hetrocyclic compounds like benzene, phenolic acids, sterols and flavonoids (Wallis, 2005). The essential oil content of the seed is usually less than 1%, whereas the fatty oil content is about 25%. The level of fatty oil and essential oil taken from different part of the coriander plant varies depending upon their origin. Petroselinic acid, linoleic acid, oleic acid, and palmitic acid are the major fatty acids. The percentage of fatty oil fluctuates from 9.9% and 27.7%. Whereas stearic acid, vaccenic acid, and myristic acid are minor fatty oil constituents. Neutral lipids make up 94.88 percent of the total lipids in coriander seed oil, with triacylglycerols accounting 95.50 per cent, 2.05 per cent free fatty acids, and 0.57% diacylglycerols followed by phospholipids which make the remaining NLs and GL. Coriander seed oil is a good source of sterols which restrict the absorption of dietary cholesterol. The main markers of sterols found in seeds of coriander include stigmasterol, b-sitosterol, arenasterol, stigmastadienol and campsterol (Sriti et al., 2009; Sriti et al., 2010). Total sterols are expected to be in the between 36.93–51.86 mg/g. Tocols are abundant in coriander oil. γ-tocopherol is the most common tocopherol followed by δ-tocopherol and α-tocopherol. Seeds of coriander contains greater level of tocotrienol with γ-tocotrienol being the abundant followed by α-tocotrienol and δ-tocotrienol (Sriti et al., 2010).

Essential oils have grown in popularity as a source of bioactive in the recent decade, with numerous possible health advantages (Burt, 2004). Using various approaches, researchers from all over the world have attempted to investigate the chemical makeup of essential oils extracted from coriander seeds. Coriander essential oil content ranges between 0.03 and 2.6 percent. Linalool, α-pinene, ɣ-terpinene, geranyl acetate, terpinolene and Geraniol are the prime ingredients of coriander fruit essential oils (Rehman et al., 2001).

The genetic origin influence the composition of the volatile oil, odour and flavour character of coriander. Linoleic acids are classified as polyunsaturated fatty acid. Coriander supplementation changes the lipid composition of carcasses by reducing saturated fatty acids and increasing monounsaturated and polyunsaturated fatty acids (Ertas et al., 2005). Ishikawa and colleagues (2003) were the first to investigate the water-soluble ingredient of the seeds. The chemicals extracted from the water-soluble part of coriander seeds included monoterpenoid glycosides, monoterpenoid glucoside sulphates, aromatic glycosides and 2-C- methylated- erythritol (Kitajima et al., 2003). Ganesan et al. (2013) reported that coriander is rich in calcium, phosphorus, magnesium, salt, zinc, and folate. Its leaves, like many other green leafy vegetables, are high in vitamins, minerals, and iron. The vitamin C concentration is plentiful among the numerous ingredients.

Conclusion

Interest in treatment of human diseases by diet-derived natural compounds is increasing day by day, hence in this context, coriander may be an ideal plant. Coriander is considered as an important ingredient of food and has been traditionally practiced all around the globe. It is the most relevant essential oil crop and its leafy vegetable use is highly popular. Nutrients like fat, proteins, vitamins, minerals etc are present in leaves and fruits and are highly fragrant. Therefore, coriander is processed to improve its flavor, productivity and increase international trade. Coriander usage as a therapeutic agent over numerous generations has recommended that it may have several impacts that are useful to humans. Coriander also exhibits several biological activities ranging from antibacterial to anticancer activities. Coriander is known as "herb of happiness" due to its versatile uses and therapeutic action against various disorders. Ample opportunities are offered by the enormous genetic diversity of the species, to develop genotypes which are adapted to particular usage and climatic conditions.

References

Abate, Limenew, Archana Bachheti, Rakesh Kumar Bachheti, Azamal Husen, Mesfin Getachew, and D. P. Pandey. "Potential Role of Forest-Based Plants in Essential Oil

Production: An Approach to Cosmetic and Personal Health Care Applications." In *Non-Timber Forest Products*, pp. 1-18. Springer, Cham, 2021.

Abdella, Ahmed, Bhagwan Singh Chandravanshi, and Weldegebriel Yohannes. "Levels of selected metals in coriander (*Coriandrum sativum* L.) leaves cultivated in four different areas of Ethiopia." *Chem. Int* 4 (2018): 189-197.

Ahmad, Taufiq, Syed Tanveer Shah, Farid Ullah, Fazal Ghafoor, and Umair Anwar. "Effect of organic fertilizer on growth and yield of coriander." *Int. J. Agri and Env. Res* 3, no. 1 (2017): 116-120.

Anon. (2009). "Coriandri fructus, in European Pharmacopoeia, 6th edn (PhEur 6.0)." European Pharmacopoeial Commission, Strasbourg; European Directorate for the Quality of Medicines (EDQM).

Bairwa, R. K., B. L. Dhaka, M. K. Poonia, B. L. Nagar, and C. M. Balai. "Coriander a potential seed spice crop of humid south eastern plains-zone of Rajasthan, India." *Int J Curr Microbiol Appl Sci* 6, no. 4 (2017): 2385-2391.

Bakkali, Fadil, Simone Averbeck, Dietrich Averbeck, and Mouhamed Idaomar. "Biological effects of essential oils–a review." *Food and chemical toxicology* 46, no. 2 (2008): 446-475.

Balasubramanian, S., P. Roselin, K. K. Singh, John Zachariah, and S. N. Saxena. "Postharvest processing and benefits of black pepper, coriander, cinnamon, fenugreek, and turmeric spices." *Critical reviews in food science and nutrition* 56, no. 10 (2016): 1585-1607.

Beyzi, Erman, Kevser Karaman, Adem Gunes, and Selma Buyukkilic Beyzi. "Change in some biochemical and bioactive properties and essential oil composition of coriander seed (*Coriandrum sativum* L.) varieties from Turkey." *Industrial Crops and Products* 109 (2017): 74-78.

Burt, Sara. "Essential oils: their antibacterial properties and potential applications in foods—a review." *International journal of food microbiology* 94, no. 3 (2004): 223-253.

Carrubba, Alessandra. "Nitrogen fertilisation in coriander (*Coriandrum sativum* L.): a review and meta-analysis." *Journal of the Science of Food and Agriculture* 89, no. 6 (2009): 921-926.

Ceska, O., S. K. Chaudhary, P. Warrington, M. J. Ashwood-Smith, G. W. Bushnell, and G. A. Poultont. "Coriandrin, a novel highly photoactive compound isolated from *Coriandrum sativum*." *Phytochemistry* 27, no. 7 (1988): 2083-2087.

Christaki, Efterpi, Eleftherios Bonos, Ilias Giannenas, and Panagiota Florou-Paneri. "Aromatic plants as a source of bioactive compounds." *Agriculture* 2, no. 3 (2012): 228-243.

Daly, Trevor, Marvin A. Jiwan, Nora M. O'Brien, and S. Aisling Aherne. "Carotenoid content of commonly consumed herbs and assessment of their bioaccessibility using an in vitro digestion model." *Plant Foods for Human Nutrition* 65, no. 2 (2010): 164-169.

Diederichsen, A. "Coriander (*Coriandrum sativum* L.). promoting the conservation and use of underutilized and negleted crops. 3. Rome: Institute of Plant Genetics and Crop Plant Research, Gatersleben." *International Plant Genetic Resources Institute* (1996): 83.

Draelos, Zoe D., Joseph Fowler, Walter G. Larsen, Sidney Hornby, Russel M. Walters, and Yohini Appa. "Tolerance of fragranced and fragrance-free facial cleansers in adults with clinically sensitive skin." *Cutis* 96, no. 4 (2015): 269-274.

Ertas, O. N., T. Guler, M. Ciftci, B. Dalkilic, and O. Yilmaz. "The effect of a dietary supplement coriander seeds on the fatty acid composition of breast muscle in Japanese quail." *Revue de médecine vétérinaire* 156, no. 10 (2005): 514.

Ganesan, P., A. Phaiphan, Y. Murugan, and B. S. Baharin. "Comparative study of bioactive compounds in curry and coriander leaves: An update." *J Chem Pharm Res* 5, no. 11 (2013): 590-594.

Grieve, M. (1971). Mandrake. *A Modern Herbal,' Dover publication INC, New York, 2.*

Ishikawa, Toru, Kyoko Kondo, and Junichi Kitajima. "Water-soluble constituents of coriander." *Chemical and Pharmaceutical Bulletin* 51, no. 1 (2003): 32-39.

Jan, Ibadullah, Muhammad Sajid, Abrar Hussain Shah, Abdur Rab, Noor Habib Khan, F. I. Wahid, A. Rahman, R. Alam, and H. Alam. "Response of seed yield of coriander to phosphorus and row spacing." *Sarhad J. Agric* 27, no. 4 (2011): 549-552.

Katiyar, R. S., R. C. Nainwal, D. Sigh, V. Chaturvedi, and S. K. Tiwari. "Effect of spacing and varieties on growth and yield of coriander (*Coriandrum sativum* L.) on reclaimed sodic waste soil." *Progressive Research* 9 (2014): 811-814.

Kassahun, Beemnet Mengesha. "Variation in pheno-qualitative characters of Ethiopian coriander (*Coriandrum sativum* L.) accessions." *Acad. Res. J. Agri. Sci. Res* 6, no. 9 (2018): 531-538.

Kassahun, Beemnet Mengesha. "Unleashing the exploitation of coriander (Coriander sativum L.) for biological, industrial and pharmaceutical applications." *Academic Research Journal of Agricultural Science and Research* 8, no. 6 (2020): 552-564.

Khan IU, Dubey W, Gupta V. Taconomical Aspect of Coriander (*Coriandrum sativum* L.). Int. J. Curr. Res., 6(11): 9926–9930.

Khan, I. U., W. Dubey, and V. Gupta. "Taxonomical aspects of Coriander (*Coriandrum sativum* L.)." In *International*, pp. 9926-9930. 2014.

Kitajima, Junichi, Toru Ishikawa, Eiko Fujimatu, Kyoko Kondho, and Tomomi Takayanagi. "Glycosides of 2-C-methyl-D-erythritol from the fruits of anise, coriander and cumin." *Phytochemistry* 62, no. 1 (2003): 115-120.

Laribi, Bochra, Karima Kouki, Mahmoud M'Hamdi, and Taoufik Bettaieb. "Coriander (*Coriandrum sativum* L.) and its bioactive constituents." *Fitoterapia* 103 (2015): 9-26.

Mandal, Shyamapada, and Manisha Mandal. "Coriander (*Coriandrum sativum* L.) essential oil: Chemistry and biological activity." *Asian Pacific Journal of Tropical Biomedicine* 5, no. 6 (2015): 421-428.

Maroufi, Kasra, Hossein Aliabadi Farahani, and Hossein Hassanpour Darvishi. "Importance of coriander (*Coriandrum sativum* L.) between the medicinal and aromatic plants." *Advances in Environmental Biology* 4, no. 3 (2010): 433-436.

Meena, K. C., D. K. Singh, I. N. Gupta, B. Singh, and S. S. Meena. "Popularazation of coriander production technologies through front line demonstrations in Hadauti region of Rajasthan." *Int. J. Seed Spices* 6, no. 2 (2016): 24-29.

Meena, S. K., N. L. Jat, B. Sharma, and V. S. Meena. "Effect of plant growth regulators and sulphur on productivity of coriander (*Coriandrum sativum* L.) in Rajasthan." *J. Environ. Sci. Int* 6 (2014): 69-73.

Mengesha, Beemnet, and G. Getinetalemaw. "Variability in Ethiopian coriander accessions for agronomic and quality traits." *African Crop Science Journal* 18, no. 2 (2010).

Mhemdi, Houcine, Elisabeth Rodier, Nabil Kechaou, and Jacques Fages. "A supercritical tuneable process for the selective extraction of fats and essential oil from coriander seeds." *Journal of Food Engineering* 105, no. 4 (2011): 609-616.

Moniruzzaman, M., M. M. Rahman, M. M. Hossain, AJM Sirajul Karim, and Q. A. Khaliq. "Effect of seed rate and sowing method on foliage production of different genotypes of coriander (*Coriandrum sativum* L.)." *Bangladesh Journal of Agricultural Research* 38, no. 3 (2013): 435-445.

Opkyde, D. L. G. "Monographs on fragrans raw materials: Coriander oil." *Food Cosm Toxicol* 11 (1973): 1077.

Peter KV (2004). Handbook of Herbs and Spices, Vol. 2; Woodhead Publishing Ltd.: Abnigton Hall, Cambridge; p. 158–174.

Puthusseri, Bijesh, Peethambaran Divya, Veeresh Lokesh, and Bhagyalakshmi Neelwarne. "Enhancement of folate content and its stability using food grade elicitors in coriander (*Coriandrum sativum* L.)." *Plant foods for human nutrition* 67, no. 2 (2012): 162-170.

Puthusseri, Bijesh, Peethambaran Divya, Veeresh Lokesh, and Bhagyalakshmi Neelwarne. "Salicylic acid-induced elicitation of folates in coriander (*Coriandrum sativum* L.) improves bioaccessibility and reduces pro-oxidant status." *Food chemistry* 136, no. 2 (2013): 569-575.

Ravi, Ramasamy, Maya Prakash, and K. Keshava Bhat. "Aroma characterization of coriander (*Coriandrum sativum* L.) oil samples." *European Food research and technology* 225, no. 3 (2007): 367-374.

Rehman, Rafia, Muhammad Asif Hanif, Zahid Mushtaq, and Abdullah Mohammed Al-Sadi. "Biosynthesis of essential oils in aromatic plants: A review." *Food Reviews International* 32, no. 2 (2016): 117-160.

Sahib, Najla Gooda, Farooq Anwar, Anwarul-Hassan Gilani, Azizah Abdul Hamid, Nazamid Saari, and Khalid M. Alkharfy. "Coriander (*Coriandrum sativum* L.): A potential source of high-value components for functional foods and nutraceuticals-A review." *Phytotherapy Research* 27, no. 10 (2013): 1439-1456.

Sharangi, A. B., and A. Roychowdhury. "Phenology and yield of coriander as influenced by sowing dates and irrigation." *The Bioscan* 9, no. 4 (2014): 1513-1520.

Sharma, Ajay, I. S. Naruka, and R. P. S. Shaktawat. "Effect of row spacing and nitrogen on growth and yield of coriander (*Coriandrum sativum* L.)." *Journal of Krishi Vigyan* 5, no. 1 (2016): 49-53.

Sharma, M. M., and R. K. Sharma. "Coriander." In: *Hand book of herbs and spices*, Vol. 2, (Ed. K. V. Peter). Woodhead Publishing Limited, Cambridge, England, 2004.

Sharma, M. M., and R. K. Sharma. "Coriander." In *Handbook of herbs and spices*, pp. 216-249. Woodhead Publishing, 2012.

Shashidhar, M. G., TP Krishna Murthy, K. Ghiwari Girish, and B. Manohar. "Grinding of coriander seeds: modeling of particle size distribution and energy studies." *Particulate Science and Technology* 31, no. 5 (2013): 449-457.

Shukla, Sanjeev, and Sanjay Gupta. "Coriander." In *Molecular targets and therapeutic uses of spices: modern uses for ancient medicine*, pp. 149-171. 2009.

Singletary, Keith. "Coriander: overview of potential health benefits." *Nutrition today* 51, no. 3 (2016): 151-161.

Small, E. "Coriander, culinary herbs." *NRC Research, Ottawa* (1997).

Smallfield, Bruce M., John W. van Klink, Nigel B. Perry, and Kenneth G. Dodds. "Coriander spice oil: effects of fruit crushing and distillation time on yield and composition." *Journal of Agricultural and Food Chemistry* 49, no. 1 (2001): 118-123.

Sriti, Jazia, Thierry Talou, Wissem Aidi Wannes, Muriel Cerny, and Brahim Marzouk. "Essential oil, fatty acid and sterol composition of Tunisian coriander fruit different parts." *Journal of the Science of Food and Agriculture* 89, no. 10 (2009): 1659-1664.

Sriti, Jazia, Wissem Aidi Wannes, Thierry Talou, Gerard Vilarem, and Brahim Marzouk. "Chemical composition and antioxidant activities of Tunisian and Canadian coriander (*Coriandrum sativum* L.) fruit." *Journal of Essential Oil Research* 23, no. 4 (2011): 7-15.

Verma, A., S. N. Pandeya, Sanjay Kumar Yadav, Styawan Singh, and Pankaj Soni. "A review on *Coriandrum sativum* (Linn.): An ayurvedic medicinal herb of happiness." *J Adv Pharm Healthc Res* 1, no. 3 (2011): 28-48.

Vaidya, V. M., and M. Gogte. "Ayurvedic Pharmacology & Therapeutic uses of medicinal plants." *Swami Prakashananda ayurvedic research center, Mumbai* (2000): 656-657.

Wallis TE (2005). Textbook of Pharmacognosy; 5th edn, S. K. Jain for CBS publishers and distibuters; New Delhi (India). pp. 125-126, 246- 248.

Yadav, S. S., I. Choudhary, L. R. Yadav, and O. P. Sharma. "Weed management in coriander (*Coriandrum sativum* L.) at varying levels of nitrogen." Journal of Spices & Aromatic Crops 25, no. 1 (2016).

Zamora, Valentin Ruben O., Manasses M. da Silva, Geronimo F. da Silva, José Amilton Santos, Dimas Menezes, and Sirleide Maria de Menezes. "Pulse drip irrigation and fertigation water depths in the water relations of coriander." *Horticultura brasileira* 37 (2019): 22-28.

Chapter 2

Botanical Characteristics and Economic Importance of *Coriandrum sativum* L.

Ankur Adhikari*

Department of Biochemistry, College of Basic Sciences and Humanities,
G. B. Pant University of Agriculture & Technology, Pantnagar, U.S. Nagar, India

Abstract

The Apiaceae family coriander (*Coriandrum sativum* L.) is amongst the most commonly exploited medicinal plants, with both nutritive and therapeutic qualities. It is one of the most prominent spices in culinary culture globally and has been known since earlier civilizations for its therapeutic benefits. It is a cross-pollinated annual aromatic herb. Its seed and leaf essential oil has since been studied extensively for their biological activity and chemical composition, including antioxidant, anti-inflammatory, hypoglycemic, anticonvulsant, hypolipidemic, anxiolytic, antibacterial, analgesic, and anti-cancer properties. Essential oils, tannins, flavonoids, reducing sugars, alkaloids, sterols, terpenoids, phenolics, fatty acids, and glycosides are all found in *Coriandrum sativum* L. Linalool, the main chemical present in seeds is known for its ability to influence several critical disease pathogenesis pathways. Proteins, carbohydrates, lipids, fibers, and a vast spectrum of trace elements, vitamins, and minerals were present substantially. Even though Coriander is said to have a wide range of traditional therapeutic applications. In summary, the information given here will aid in raising awareness of this medicinal species by describing its botanical, economic, new pharmacological, and clinical uses. This book chapter covered a wide range of topics connected to the botanical, traditional, and

* Corresponding Author's Email: ankuradk@gmail.com.

In: *Coriandrum sativum*: Origin, Uses and Nutrition
Editors: Mamta Pujari, Bharat Kapoor, Neeta Pande et al.
ISBN: 978-1-68507-842-3

medicinal properties of coriander, indicating its potential as a functional food for improving the quality of life in the period of aging and lifestyle-associated illnesses.

Keywords: coriander, botanical, linalool, medicinal, gastrointestinal

1. Introduction

Herbal medicine is the world's ancient type of medication. It was the lifeblood of several ancient civilizations and is remains the most frequently used type of medicine today (Al-Snafi and Ali, 2016). Traditional medicinal plants are becoming more common but are indeed commonly used. This has piqued the desire of many scholars, who have been urged to examine medicinal plants to investigate the biological activities of their bioactive components. Medicinal plants have many biological activities of their biochemical components (Seidemann, 2005). Coriander has been used since approximately 1550 BC, making it one of the world's ancient spices. The crops' seeds are consumed as a spice, while the delicate green leaves are consumed in cooking. In the Rabi season, it has exhibited to be a lucrative crop.

Coriandrum sativum L. is one of the intriguing annual herbaceous medicinal plants with a diploid chromosome (2n = 22). Its seeds are commercially utilized as a condiment in many areas of the world or as a principal element of Indian curry food (Laribi et al., 2015). Numerous ingredients, such as fish and steak, bakery, and confectionery items, are flavored with seeds. It was discovered that Coriander has many drug-related effects such as anti-proliferative (Nakano et al., 1998), anti-hyperglycemic (Eidi et al., 2009), hypotensive, anti-hyperlipidemia (Chithra and Leelamma, 1999), anti-perpetuity, digestive stimulants (Platel et al., 2000), anti-fertility (Al-Said et al., 1987), and hypotensive (Burdock et al., 2009).

While natural antioxidants of Coriander have been extensively studied for their biological activities and chemical properties. This chapter aims to emphasize the significance of coriander as a root source of economically important components and outline their bioactivities, including the pharmacological studies that have been conducted on this species, with a thorough overview of the gaps and prospective research opportunities.

2. Botanical Feature of *Coriandrum sativum* L.

Coriandrum sativum L., a Mediterranean perennial herb of the Apiaceae family, also known as the carrot family, is widely produced in India, Russia, Morocco, and central Europe, and Bangladesh has been nurtured from human civilization (Small, 1997). It has a wide range of adaptability as a crop across the globe, thriving well in various soils and climates (Perseglove et al., 1981; Simon, 1990). Furthermore, most coriander varieties have a short life cycle, allowing farmers to plant during any period of the cropping season in nearly any area. The green leaves of coriander are known as Asotu and cilantro in Eastern Anatolia and the United state, respectively. It is named differently in various regions of the world, such as in English (Coriander, cilantro), Hindi (Dhaniya), Sanskrit (Dhanayaka, kusthumbari), Japanese (Koendoro), German (Koriander, Wanzendill, Schwindelkorn), French (Coriander, Coriandre cultivé), Portuguese (Coentro, Coriandro), Spanish (Cilantro, Cilandrio, Coriandro,), Swedish (Coriander), Chinese (Hu sui, Yuan sui), Arabic (Kuzbara, kuzbura) (Verma et al., 2011). This perennial spice herb with cross-pollination reaches up to 50 cm and is endemic to the Mediterranean and Near East. European Union former Members, Hungary, Czech Romania, Slovenia, Poland, Morocco, Canada, India, Pakistan, Iran, Turkey, Guatemala, Mexico, and Argentina have been the leading commercial coriander grower of recent years (Kiehn, 1992).

The climatic condition for the cultivation of Coriander requires dry heat for the production of high-quality fruits. Coriander has complex umbels in its inflorescence, which are a hallmark of the family. The major umbel is the earliest to blossom and develops as it ends the plant's principal stem. The majority of coriander genotypes generate numerous higher-order side branches, all of which culminate in an inflorescence. The peripheral flowers are the first to blossom in an umbel and also in the umbellets. The peripheral flora in a umbellet is protandrous, i.e., the anthers are fertile and begin to dump pollen two or three days before the same flower is stigmatized. There are five petals on each floret. The petals of a umbellet's marginal florets, especially those near the umbel's margin, are skewed, with the petals pointing to the outside of the umbel being larger. Coriander's floral characteristics allow for a high proportion of outcrossing. On the other hand, Coriander is not an obligatory outcrosser, and there is a lot of self-pollination on geitonogamy. It attracts various insects, including honeybees, and their pollination enhances seed set and fruit production (Martins, 2018; Patil and Pastagia, 2016). However, incomplete pollination can't generate fruits by some blooms. It might occur that none or just one of the two seeds in fruit (*Schizocarpium*) typically develops, which affects yield and reduces fruit quality.

***Coriandrum sativum* L. Taxonomic classification**

Kingdom	-	Plantae
Subkingdom	-	Tracheobionta
Superdivison	-	Spermatophyta
Divison	-	Magnoliophyta
Class	-	Magnoliopsida
Subclass	-	Rosidae
Order	-	Apiales
Family	-	Apiaceae
Genus	-	Coriandrum L.
Species	-	*Coriandrum sativum* L.

3. Economical Aspects

Coriander leaves are widely used in salads and as a delicacy in several parts of the world. Coriander is cultivated chiefly as a yearly summer, but in southern Russia and Turkey, it is cultivated as a winter annual (Unlukara et al., 2016). In various regions from the brief circumpolar period Alaska of the United States to tropical Kenya, small-scale coriander cultivation for fruit and vegetable usage may be seen (Martins, 2018). Resistance to a variety of ailments is crucial for all forms of coriander use. Coriander is resistant to *Protomyces macrosporus* Unger (stem gall disease). It is resistant to stem gall disease in India, where the infection is economically essential (Singh et al., 2003). Blossom blight, or wilting of the umbels, occurs during or immediately after bloom and is affected by a range of diseases, along with the bacterium *Pseudomonas syringae* pv. Coriandricola and *Ramularia coriandri* fungus downgrade the yield and quality of fruit (Cerkauskas, 2009). Plant breeding in Russia focused on *Ramularia* resistance (Bochkarev et al., 2014).

4. Phytochemical Components

According to phytochemical analysis, *Coriandrum sativum* L. contains phenolics, flavonoids, alkaloids, fatty acids, essential oil, tannins, terpenoids, sugars, sterols, and glycosides. It contains protein (12.37-21.93 g), total lipid

(4.78-17.77 g), carbohydrate (52.10-54.99 g), fiber (10.40-41.9 g), Ca (709-1246 mg), Fe (16.32-42.46 mg), P (409-481 mg), Mg (330-694 mg), K (4466-1267 mg), Na (35-211 mg), Zn (4.72 mg), Vit-C (21-566.7 mg), Thiamin (0.239-1.252 mg), Riboflavin (0.290-1.5 mg), Niacin (2.130-10.7 mg), Vit-B (0-120 μg) respectively (Coşkuner and Erşan, 2007; Bhat et al., 2014; Diederichsen, 1996). The essential oil of coriander fruits has been researched in many parts of the world and reported to differ from one another. It is published that coriander seed oil is said to contain ~70% linalool and hydrocarbons (20%), and the coriander leaf oil's constitution is entirely different from the seed oil's (Guenther, 1950). In seed oil of Coriander Rastogi and Mehrotra, (1993) detected α-pinene, limonene, β-phellandrene, eucalyptol, linalool, borneol, β-caryophyllene, citronellol, geraniol, thymol, linalyl acetate, geranyl acetate, caryophyllene oxide, elemol, and methyl heptenol by TLC. It was observed that the essential oil concentration in mature fruits is deficient (<1%); the oil comprises mostly of linalool (50-60%) and ~20% terpenes (Telci et al., 2006).

Bhuiyan et al., (2009) studied the physiochemical properties and components of *Coriandrum sativum* L. leaves and seeds by gas chromatography. Linalool is the most abundant volatile component in seeds, accounting for more than half of the essential oil. In coriander seeds, fatty acids are the essential components such as petroselenic, linoleic, palmitic, and stearic acids (Ramadan and Mörsel, 2003). Additionally, breakage of the unique double bond in petroselenic acid results in the formation of lauric acid, which is employed to make surfactants and food goods, and adipic acid, which is used to make nylon (Isbell et al., 2006). Makeeva et al. (2014) identified Linalool, Nerol, Bornel, Gereniol, and α-Terpinene as principal components in various Coriander dried fruits' oil. The essential and fatty oils were the most significant components of coriander fruits, and it varies from 0.03-2.6% and 9.9-27.7%, respectively (Coşkuner and Erşan, 2007).

5. Biological Activities

Coriander fruits and herbs are extensively used in conventional therapy and have various benefits on the gastrointestinal tract and immune system, as evidenced by current studies (Nadeem et al., 2013; Prachayasittikul et al., 2018). It is confirmed that the leaves and fruits of Coriander have antioxidants and have antifungal and antimicrobial effects (Silva and Domingues, 2017). Owing to the loss of antibiotic capability as antimicrobial agents due to drug resistance, scholars' attention has moved to plant-derived products, which has resulted in a surge in the adoption of plant extracts in traditional medicine for antibacterial properties. Coriander has been found to have antibacterial properties towards various organisms (Laribi et al., 2015; Jiang et al., 2015). A unique antimicrobial peptide Plantaricin CS has a significant antibacterial property against gram-negative bacteria extracted from Coriander leaf extract (Zare-Zardini et al., 2012). Bezalwar et al., (2019) investigated multidrug resistance against *Pseudomonas aeruginosa* by the synergistic action of eight different antibiotics and *Coriandrum sativum* L. extract. Common biochemical pathway inhibition, enzyme degradation, and cell wall-active substances are some of the pathways used by the synergism of medication and extract combinations.

Plants generate many secondary compounds, which are biosynthetically produced from primary metabolites and constitute a significant root of pharmaceutical medicines. The aqueous extracts of *Coriander sativum* L. depicted anxiolytic activity by increasing the time spent on open arms and the percentage of open arm entries, compared to the control group in various studies (Latha et al., 2015; Pathan et al., 2011;). *Coriandrum sativum* L. extracts exhibited anti-anxiety repercussions that were virtually identical to diazepam at 100 and 200 mg/kg doses but did not induce anti-anxiety activity in all animals at 50 mg/kg dosage (Mahendra et al., 2011). However, interaction with the adrenergic, dopaminergic, and GABA-ergic systems, diethyl ether of seeds of *Coriandrum sativum* L. demonstrated a more substantial antipsychotic than aqueous extract (Sudha et al., 2011). The essential oil enhanced the pentobarbital-induced sleeping time (Emam and Hamedani, 2006). The findings suggested that *Coriandrum sativum* L. had a sleep-inducing effect without being neurotoxic (Rakhshandeh et al., 2012). Consequently, there were substantial antioxidant and anticonvulsant properties in the hydroalcoholic extract of *Coriandrum sativum* L. aerial portions (Rakhshandeh et al., 2012).

According to the findings, *Coriandrum sativum* L. with water-soluble compound(s) can produce neuroprotective action. In contrast, some of its components can be cytotoxic under stressed situations like hypoglycemia (Ghorbani et al., 2011). There are several other properties of *Coriandrum sativum* L. such as antibacterial (Oudah, 2010), antifungal (Soares et al., 2012), anthelmintic (Khani et al., 2012) and insecticidal effects (Benelli et al., 2013), antioxidant effect (de Almeida Melo et al., 2005; Wong and David, 2006; Hashim et al., 2005; Wangensteen et al., 2004), hypolipidemic effect (Kousar et al., 2011; Joshi et al., 2012), anti-inflammatory and analgesic effects (Nair et al., 2013; Haj et al., 2003), antidiabetic effect (Waheed et al., 2006; Naquvi et al., 2012), mutagenic and antimutagenic effect (Reyes et al., 2010; Cortés-Eslava et al., 2004), anticancer effect (Gomez-Flores et al., 2010; Tang et al., 2013), cardiovascular effects (Medhin et al., 1986; Patel et al., 2012), gastrointestinal effects (Al-Mofleh et al., 2006; Zaidi et al., 2012), hepatoprotective effect (Sreelatha et al., 2009; Pandey et al., 2011; Samojlik et al., 2010), dermatological effect (Hwang et al., 2014; Park et al., 2014), effect on fertility (Al-Said et al., 1987).

6. Uses

The two primary items derived from coriander plants are a fresh green herb and a dried spice. Although all parts of this plant are eatable, but mature seed and fresh herbage have entirely distinct odors and flavors. In the folk medicine systems in distinctive cultures, Coriander has been extensively exploited as a cooking element and traditional medicines to prevent multiple disorders (Sahib et al., 2013; Ugulu et al., 2009). So, when the plant is young, the complete plant is utilized to make chutneys, sauces, curries, and soups. Curry powder, pickling spices, sausages, and seasonings all employ coriander fruits as condiments. Coriander roots have a deeper, stronger flavor than leaves, and it is utilized chiefly in various Asian cuisines (e.g., soups and stews) (Verma et al., 2011).

Cilantro is also employed in Indian traditional medicine to treat intestinal, respiratory, and urinary system diseases since it has diaphoretic, diuretic, carminative, and muscle relaxant properties. The plant Coriander is used as a traditional medicine in northern parts of Pakistan to cure diarrhea, stomach problems, vomiting, dysentery, jaundice, and flatulence (Khan and Khatoon, 2008). In addition, Coriander has become one of the conventional medicinal herbs for the treatment of hyperglycemia. Therefore, an infusion of seeds of Coriander as an antidiabetic agent is employed in some countries like Jordan, Saudi Arabia, and Morocco (Al-Rowais and Norah, 2002; Otoom et al., 2006; Tahraoui et al., 2007). Lawrence (1992) lists coriander seed oil as one of the world's top 20 essential oils, and its economic worth is determined by its physical characteristics, chemical makeup, and fragrance (Smallfield et al., 2001). The young leaf has a distinct fragrance and is commonly employed as an essential spice in Thai and Vietnamese cuisine (Gil et al., 2002). In addition, the frequent utilization of a coriander seed decoction was stated to reduce blood lipid levels in Ayurvedic literature successfully. Furthermore, Coriander has been documented in Moroccan and Palestinian pharmacopeia as a typical diuretic or to heal urinary infections (Eddouks et al., 2002; Abu-Rabia, 2005). Sometimes, it is employed as a flavoring substance in bakery products. Its gentle, pleasant, somewhat spicy smell combines well with oriental perfumes. The leaf and seed oils include significant concentrations of linalool and 2-decenoic acid, making them extremely valuable in medicine and fragrance. Linalool is used as a chemical intermediary since its downstream result is vitamin E; it is used as a fragrance in cosmetics goods such as shampoo, shop, detergent, and lotion. The coriander seeds are exploited in the fragrance and flavors sector to make steam distilled essential oils and solvent extracted oleoresin (Purseglove et al., 1981). Seed essential oil is used in meditation, perfumes, soap manufacturing, and food flavoring, and it also possesses antifungal and antibacterial properties (Agri-facts, 1998). Seeds, on the other hand, were administered topically to relieve swelling and inflammation. Externally, dried green Coriander was employed to facilitate the blistering feeling and pain associated with erysipelas and lymphadenopathy. Cilantro drops in the nose serve as a hemostat, preventing epistaxis. The leaves' fresh juice has traditionally been utilized as a mouthwash for sore throats and stomatitis. For swellings and boils, a paste of leaves was administered locally, and for headaches, it was rubbed across the forehead (Pandey, 2010).

Conclusion

Since ancient times, *Coriandrum sativum* L. has been exploited in humankind as food, scent, and medication. It showed a wide spectrum of biological activities that are advantageous for aging illnesses and disorders caused by a contemporary sedentary lifestyle. Coriander is globally the most relevant essential oil crop, and leafy vegetable use is increasingly popular. According to the wide range of pharmacological activities, *Coriandrum sativum* L. should be considered a promising source of many drugs because of its safety and effectiveness. Coriander has been authorized as a dietary component, although more research into toxicities and complications at long-term therapeutic doses is needed.

References

Abu-Rabia, A. 2005. Herbs as a food and medicine source in Palestine. *Asian Pacific Journal of Cancer Prevention*, 6(3), 404.

Agri-facts, Coriander. "*Practical information for Alber-ta's agriculture industry.*" Accessed online at: http://agric.gov.ab.ca/agdex/100/147_20-2. tml# top (1998).

Al-Mofleh, I. A., A. A. Alhaider, J. S. Mossa, M. O. Al-Sohaibani, S. Rafatullah, and S. Qureshi. "Protection of gastric mucosal damage by *Coriandrum sativum* L. pretreatment in Wistar albino rats." *Environmental Toxicology and Pharmacology* 22, no. 1 (2006): 64-69.

Al-Rowais, N. A. "Herbal medicine in the treatment of diabetes mellitus." *Saudi medical journal* 23, no. 11 (2002): 1327-1331.

Al-Said, M. S., K. I. Al-Khamis, Mohammad W. Islam, N. S. Parmar, M. Tariq, and A. M. Ageel. "Post-coital antifertility activity of the seeds of *Coriandrum sativum* L. in rats." *Journal of ethnopharmacology* 21, no. 2 (1987): 165-173.

Al-Snafi, A. E. "A review on chemical constituents and pharmacological activities of *Coriandrum sativum.*" *IOSR Journal of Pharmacy* 6, no. 7 (2016): 17-42.

Benelli, G., Guido Flamini, Giulia Fiore, Pier Luigi Cioni, and Barbara Conti. "Larvicidal and repellent activity of the essential oil of *Coriandrum sativum* L. (Apiaceae) fruits against the filariasis vector Aedes albopictus Skuse (Diptera: Culicidae)." *Parasitology research* 112, no. 3 (2013): 1155-1161.

Bezalwar, P. M., and Vijay N. Charde. "Study on synergistic action of *Coriandrum sativum* L. seed extracts on antibiotics against multidrug resistant *P. aeruginosa.*" *Environment Conservation Journal* 20, no. 3 (2019): 83-88.

Bhat, S., Kaushal, P., Kaur, M., & Sharma, H. K. (2014). "Coriander (*Coriandrum sativum* L.): Processing, nutritional and functional aspects." *African Journal of plant science*, 8(1), 25-33.

Bhuiyan, M. N. I., Jaripa Begum, and Mahbuba Sultana. "Chemical composition of leaf and seed essential oil of *Coriandrum sativum* L. from Bangladesh." ||| *Bangladesh Journal of Pharmacology*||| 4, no. 2 (2009): 150-153.

Bochkarev, N. I., Zelentsov C. V., Moshenko E. V. (2014). "*Morfologija, taksonomija, metodu selekcii I characteristika sortov koriandra posevnovo (obzor): Maclichnye kul'tury* [*Morphology, taxonomy, selection method And characteristics of coriander varieties sowing (review): Oil crops*]." 159-160: 178-195.

Burdock, G. A., and Ioana G. Carabin. (2009). "Safety assessment of coriander (*Coriandrum sativum* L.) essential oil as a food ingredient." *Food and Chemical Toxicology* 47, 1: 22-34.

Cerkauskas, R. F. (2009). "Bacterial leaf spot of cilantro (*Coriandrum sativum*) in Ontario." *Canadian journal of plant pathology* 31, 1: 16-21.

Chithra, V., and S. Leelamma. "*Coriandrum sativum L. changes the levels of lipid peroxides and activity of antioxidant enzymes in experimental animals.*" (1999).

Cortés-Eslava, J., Sandra Gómez-Arroyo, Rafael Villalobos-Pietrini, and Jesús Javier Espinosa-Aguirre. "Antimutagenicity of coriander (*Coriandrum sativum*) juice on the mutagenesis produced by plant metabolites of aromatic amines." *Toxicology letters* 153, no. 2 (2004): 283-292.

Coşkuner, Y., and Erşan Karababa. "Physical properties of coriander seeds (*Coriandrum sativum* L.)." *Journal of Food Engineering* 80, no. 2 (2007): 408-416.

de Almeida Melo, Enayde, Jorge Mancini Filho, and Nonete Barbosa Guerra. "Characterization of antioxidant compounds in aqueous coriander extract (*Coriandrum sativum* L.)." *LWT-Food Science and Technology* 38, no. 1 (2005): 15-19.

Diederichsen, A. *Coriander: Coriandrum Sativum* L. Vol. 3. Bioversity International, 1996.

Eddouks, M., Maghrani, M., Lemhadri, A., Ouahidi, M. L., & Jouad, H. (2002). Ethnopharmacological survey of medicinal plants used for the treatment of diabetes mellitus, hypertension and cardiac diseases in the south-east region of Morocco (Tafilalet). *Journal of ethnopharmacology*, 82(2-3), 97-103.

Eidi, M., Akram Eidi, Ali Saeidi, Saadat Molanaei, Alireza Sadeghipour, Massih Bahar, and Kamal Bahar. "Effect of coriander seed (*Coriandrum sativum* L.) ethanol extract on insulin release from pancreatic beta cells in streptozotocin-induced diabetic rats." *Phytotherapy Research: An International Journal Devoted to Pharmacological and Toxicological Evaluation of Natural Product Derivatives* 23, no. 3 (2009): 404-406.

Emam, G. M., and Hamedani G. Heydari. "*Sedative-hypnotic activity of extracts and essential oil of coriander seeds.*" (2006): 22-27.

Ghorbani, A., Hassan Rakhshandeh, Elham Asadpour, and Hamid Reza Sadeghnia. "Effects of *Coriandrum sativum* L. extracts on glucose/serum deprivation-induced neuronal cell death." *Avicenna Journal of Phytomedicine* 2, no. 1 (2011): 4-9.

Gomez-Flores, R., Humberto Hernández-Martínez, Patricia Tamez-Guerra, Reyes Tamez-Guerra, Ramiro Quintanilla-Licea, Enriqueta Monreal-Cuevas, and Cristina Rodríguez-Padilla. "Antitumor and immunomodulating potential of *Coriandrum sativum*." *Journal of Natural Products* 3 (2010): 54-63.

Guenther, E. "The Essential Oils, Vol. IV." *The Essential Oils, Vol. IV.* (1950).

Haj, H. V. A., A. Ghannadi, and Badiah Sharif. "*Anti-inflammatory and analgesic effects of Coriandrum sativum L. in animal models.*" (2003): 8-15.

Hashim, M. S., S. Lincy, V. Remya, M. Teena, and L. Anila. "Effect of polyphenolic compounds from *Coriandrum sativum* on H2O2-induced oxidative stress in human lymphocytes." *Food chemistry* 92, no. 4 (2005): 653-660.

Hwang, E., Do-Gyeong Lee, Sin Hee Park, Myung Sook Oh, and Sun Yeou Kim. "Coriander leaf extract exerts antioxidant activity and protects against UVB-induced photoaging of skin by regulation of procollagen type I and MMP-1 expression." *Journal of medicinal food* 17, no. 9 (2014): 985-995.

Isbell, T. A., Lindsay A. Green, Stephanie S. DeKeyser, Linda K. Manthey, James A. Kenar, and Steven C. Cermak. "Improvement in the gas chromatographic resolution of petroselinate from oleate." *Journal of the American Oil Chemists' Society* 83, no. 5 (2006): 429-434.

Jiang, D. M., Yuan Zhu, J. N. Yu, and X. M. Xu. "Advances in research of pharmacological effects and formulation studies of linalool." *Zhongguo Zhong yao za zhi= Zhongguo zhongyao zazhi= China journal of Chinese materia medica* 40, no. 18 (2015): 3530-3533.

Joshi, S. C., Nidhi Sharma, and Preeti Sharma. "Antioxidant and lipid lowering effect of *Coraindrum sativum* in cholosterol fed rabbits." *Internat. J. Pharm. and Pharmac. Sci* 4, no. 3 (2012): 231-234.

Khan, S. W., and S. U. R. A. Y. Y. A. Khatoon. "Ethnobotanical studies on some useful herbs of Haramosh and Bugrote valleys in Gilgit, northern areas of Pakistan." *Pakistan Journal of Botany* 40, no. 1 (2008): 43.

Khani, A., and Tahere Rahdari. "Chemical composition and insecticidal activity of essential oil from *Coriandrum sativum* seeds against *Tribolium confusum* and *Callosobruchus maculatus*." *International Scholarly Research Notices* 2012 (2012).

Kiehn, F. A. and Reimer M., *Alternative crops for the prairies, Agriculture Canada Research Station,* Morden, Manitoba, 15-16 (1992).

Kousar, S., N. Jahan, Khalil-ur-Rehman, and S. Nosheen. "Antilipidemic activity of *Coriandrum sativum*." *Bioscience Research* 8, no. 1 (2011): 8-14.

Laribi, B., Karima Kouki, Mahmoud M'Hamdi, and Taoufik Bettaieb. "Coriander (*Coriandrum sativum* L.) and its bioactive constituents." *Fitoterapia* 103 (2015): 9-26.

Laribi, B., Kouki, K., M'Hamdi, M., & Bettaieb, T. (2015). Coriander (*Coriandrum sativum* L.) and its bioactive constituents. *Fitoterapia*, 103, 9-26.

Latha, K., B. Rammohan, B. P. V. Sunanda, MS Uma Maheswari, and Surapaneni Krishna Mohan. "Evaluation of anxiolytic activity of aqueous extract of *Coriandrum sativum* L. in mice: A preliminary experimental study." *Pharmacognosy research* 7, no. Suppl 1 (2015): S47.

Lawrence, B. M. "A planning scheme to evaluate new aromatic plants for the flavor and fragrance industries." *New crops* 1 (1993): 620-627.

Mahendra, P., and Shradha Bisht. "Anti-anxiety activity of *Coriandrum sativum* assessed using different experimental anxiety models." *Indian journal of pharmacology* 43, no. 5 (2011): 574.

Makeeva, D. R., E. M. Kryukova, and Elena E. Konovalova. "Tourism as preferred direction in the strategy of substitution of industry branches in mono-territories of Russian Federation." *World Applied Sciences Journal* 30, no. 1 (2014): 176-178.

Martins, D. J. Pollination in some Kenyan smallholder crops. In: Roubik D. E. (ed) *The pollination of cultivated plants, a compendium for practitioners,* vol 1, pp 121-135: (2018). Food and agriculture organization of the United Nations.

Medhin, D. G., P. Bakos, and G. Verzar-Petri. "Hypotensive effects of *Lupinus termis* and *Coriandrum sativum* in Anaesthetized Rats. A preliminary study." *Acta Pharmaceutica Hungarica* 56, no. 2 (1986): 59-63.

Nadeem, M., Faqir Muhammad Anjum, Muhammad Issa Khan, Saima Tehseen, Ahmed El-Ghorab, and Javed Iqbal Sultan. "Nutritional and medicinal aspects of coriander (*Coriandrum sativum* L.): A review." *British Food Journal* (2013).

Nair, V., Surender Singh, and Yogendra Kumar Gupta. "Anti-granuloma activity of *Coriandrum sativum* L. in experimental models." *Journal of ayurveda and integrative medicine* 4, no. 1 (2013): 13.

Nakano, Y., Hisashi Matsunaga, Tetsuya Saita, Masato Mori, Mitsuo Katano, and Hikaru Okabe. "Antiproliferative Constituents in Umbelliferae Plants II.: Screening for Polyacetylenes in Some Umbelliferae Plants, and Isolation of Panaxynol and

Falcarindiol from the Root of Heracleum moellendorffii." *Biological and Pharmaceutical Bulletin* 21, no. 3 (1998): 257-261.

Naquvi, K. J., M. O. H. D. Ali, and J. Ahmad. "Antidiabetic activity of aqueous extract of *Coriandrum sativum* L. fruits in streptozotocin induced rats." *Indian J Exp Biol* 42, no. 9 (2004): 909-12.

Otoom, S. A., S. A. Al-Safi, Z. K. Kerem, and A. Alkofahi. "The use of medicinal herbs by diabetic Jordanian patients." *Journal of herbal pharmacotherapy* 6, no. 2 (2006): 31-41.

Oudah, I. "Evaluation of aqueous and ethanolic extraction for Coriander seeds, leaves and stems and studying their antibacterial activity." *Iraqi National Journal of Nursing Specialties* 1, no. 23 (2010): 1-7.

Pandey, A., P. Bigoniya, V. Raj, and K. K. Patel. "Pharmacological screening of *Coriandrum sativum* L. for hepatoprotective activity." *Journal of Pharmacy and Bioallied sciences* 3, no. 3 (2011): 435.

Pandey, S. "*Coriandrum sativum*: a biological description and its uses in the treatment of various diseases." *International Journal of Pharmacy and Life Sciences (IJPLS)* 1, no. 3 (2010): 119-126.

Park, G., Hyo Geun Kim, Soonmin Lim, Wonil Lee, Yeomoon Sim, and Myung Sook Oh. "Coriander alleviates 2, 4-dinitrochlorobenzene-induced contact dermatitis-like skin lesions in mice." *Journal of medicinal food* 17, no. 8 (2014): 862-868.

Patel, D. K., Swati N. Desai, Hardik P. Gandhi, Ranjitsinh V. Devkar, and A. V. Ramachandran. "Cardio protective effect of *Coriandrum sativum* L. on isoproterenol induced myocardial necrosis in rats." *Food and chemical toxicology* 50, no. 9 (2012): 3120-3125.

Pathan, A. R., K. A. Kothawade, and Mohd Nadeem Logade. "Anxiolytic and analgesic effect of seeds of *Coriandrum sativum* L." *Int J Res Pharm Chem* 1 (2011): 1087-99.

Patil, P. N., and J. J. Pastagia. "Effect of bee pollination on yield of coriander, *Coriandrum sativum* L." *International Journal of Plant Protection* 9, no. 1 (2016): 79-83.

Platel, K., and K. Srinivasan. "Stimulatory influence of select spices on bile secretion in rats." *Nutrition Research* 20, no. 10 (2000): 1493-1503.

Prachayasittikul, V., Supaluk Prachayasittikul, Somsak Ruchirawat, and Virapong Prachayasittikul. "Coriander (*Coriandrum sativum*): A promising functional food toward the well-being." *Food Research International* 105 (2018): 305-323.

Purseglove, J. W., E. G. Brown, C. L. Green, and S. R. J. Robbins. "Spices. Volumes 1 and 2." Purseglove, JW; Brown, EG; Green, CL; Robbins, SRJ: *Spices. Volumes 1 and 2.* (1981).

Rakhshandeh, H., Hamid Reza Sadeghnia, and Ahmad Ghorbani. "Sleep-prolonging effect of *Coriandrum sativum* L. hydro-alcoholic extract in mice." *Natural product research* 26, no. 22 (2012): 2095-2098.

Rastogi, R. P., B. N. Mehrotra, and R. P. Pastogi. "Compendium of Indian Medicinal plants. Central drug research institute." *Lucknow and Publications & Information Directorate*, New Delhi, India 2 (1993): 10.

Reyes, M. R., Jorge Reyes-Esparza, Oscar Torres Ángeles, and Lourdes Rodríguez-Fragoso. "Mutagenicity and safety evaluation of water extract of *Coriander sativum* leaves." *Journal of food science* 75, no. 1 (2010): T6-T12.

Sahib, N. G., Anwar, F., Gilani, A. H., Hamid, A. A., Saari, N., & Alkharfy, K. M. (2013). Coriander (*Coriandrum sativum* L.): A potential source of high-value components for functional foods and nutraceuticals-A review. *Phytotherapy Research*, *27*(10), 1439-1456.

Samojlik, I., Neda Lakic, Neda Mimica-Dukic, Kornelia Đaković-Švajcer, and Biljana Bozin. "Antioxidant and hepatoprotective potential of essential oils of coriander (*Coriandrum sativum* L.) and caraway (*Carum carvi* L.) (Apiaceae)." *Journal of Agricultural and Food Chemistry* 58, no. 15 (2010): 8848-8853.

Seidemann, J. (2005). *World spice plants: economic usage, botany, taxonomy*. Springer Science & Business Media.

Silva, F., and Fernanda C. Domingues. "Antimicrobial activity of coriander oil and its effectiveness as food preservative." *Critical reviews in food science and nutrition* 57, no. 1 (2017): 35-47.

Simon, J. E. "Essential oils and culinary herbs." In *Advances in new crops. Proceedings of the first national symposium 'New crops: research, development, economics,'* Indianapolis, Indiana, USA, 23-26 October 1988., pp. 472-483. Timber Press, 1990.

Singh, H. B., A. Singh, A. Tripathi, S. K. Rai, R. S. Katiyar, J. K. Johri, and S. P. Singh. "Evaluation of Indian coriander accessions for resistance against stem gall disease." *Genetic Resources and Crop Evolution* 50, no. 4 (2003): 339-343.

Small, E. *Culinary herbs*. Ottawa, NRC Research Press, 1997, pp 219-25.

Smallfield, B. M., John W. van Klink, Nigel B. Perry, and Kenneth G. Dodds. "Coriander spice oil: effects of fruit crushing and distillation time on yield and composition." *Journal of Agricultural and Food Chemistry* 49, no. 1 (2001): 118-123.

Soares, B. V., Selene M. Morais, Raquel Oliveira dos Santos Fontenelle, Vanessa A. Queiroz, Nadja S. Vila-Nova, Christiana Pereira, Edy S. Brito et al. "Antifungal activity, toxicity and chemical composition of the essential oil of *Coriandrum sativum* L. fruits." *Molecules* 17, no. 7 (2012): 8439-8448.

Sreelatha, S., P. R. Padma, and M. Umadevi. "Protective effects of *Coriandrum sativum* extracts on carbon tetrachloride-induced hepatotoxicity in rats." *Food and Chemical Toxicology* 47, no. 4 (2009): 702-708.

Sudha, K., Gumate Deepak, Kokane Sushant, Patil Vipul, and Naikwade Nilofer. "Study of antidepressant like effect of *Coriandrum sativum* and involvement of monoaminonergic and Gabanergic system." *IJRAP* 2 (2011): 267-270.

Tahraoui, A., J. El-Hilaly, Z. H. Israili, and B. Lyoussi. "Ethnopharmacological survey of plants used in the traditional treatment of hypertension and diabetes in south-eastern Morocco (Errachidia province)." *Journal of ethnopharmacology* 110, no. 1 (2007): 105-117.

Tang, E. L. H., Jayakumar Rajarajeswaran, Shin Yee Fung, and M. S. Kanthimathi. "Antioxidant activity of *Coriandrum sativum* L. and protection against DNA damage and cancer cell migration." *BMC complementary and alternative medicine* 13, no. 1 (2013): 1-13.

Telci, I., Ozlem Gul Toncer, and Nermin Sahbaz. "Yield, essential oil content and composition of *Coriandrum sativum* varieties (var. vulgare Alef and var. microcarpum DC.) grown in two different locations." *Journal of Essential Oil Research* 18, no. 2 (2006): 189-193.

Ugulu, I., Suleyman Baslar, Nurettin Yorek, and Yunus Dogan. "The investigation and quantitative ethnobotanical evaluation of medicinal plants used around Izmir province, Turkey." *Journal of Medicinal plants research* 3, no. 5 (2009): 345-367.

Unlukara, A., Erman Beyzi, I. P. E. K. Arif, and Bilal Gurbuz. "Effects of different water applications on yield and oil contents of autumn sown coriander (*Coriandrum sativum* L.)." *Turkish Journal of Field Crops* 21, no. 2 (2016): 200-209.

Verma, A., S. N. Pandeya, Sanjay Kumar Yadav, Styawan Singh, and Pankaj Soni. "A review on *Coriandrum sativum* (Linn.): An ayurvedic medicinal herb of happiness." *J Adv Pharm Healthc Res* 1, no. 3 (2011): 28-48.

Waheed, A., G. A. Miana, S. I. Ahmad, and Munir A. Khan. "Clinical investigation of hypoglycemic effect of *Coriandrum sativum* in type-2 (NIDDM) diabetic patients." *Pakistan Journal of Pharmacology* 23, no. 1 (2006): 7-11.

Wangensteen, H., Anne Berit Samuelsen, and Karl Egil Malterud. "Antioxidant activity in extracts from coriander." *Food chemistry* 88, no. 2 (2004): 293-297.

Wong, P. Y. Y., and David D. Kitts. "Studies on the dual antioxidant and antibacterial properties of parsley (*Petroselinum crispum*) and cilantro (*Coriandrum sativum*) extracts." *Food chemistry* 97, no. 3 (2006): 505-515.

Zaidi, S. F., Jibran Sualeh Muhammad, Saeeda Shahryar, Khan Usmanghani, Anwarul-Hassan Gilani, Wasim Jafri, and Toshiro Sugiyama. "Anti-inflammatory and cytoprotective effects of selected Pakistani medicinal plants in Helicobacter pylori-infected gastric epithelial cells." *Journal of Ethnopharmacology* 141, no. 1 (2012): 403-410.

Zardini, H. Z., Behnaz Tolueinia, Zeinab Momeni, Zohre Hasani, and Masumeh Hasani. "Analysis of antibacterial and antifungal activity of crude extracts from seeds of *Coriandrum sativum* L." *Gomal journal of medical sciences* 10, no. 2 (2012).

Chapter 3

Chemical Composition, Nutritive and Nutraceutical Potential of Coriander

Milica G. Acimovic[1,*], Ana Marjanović Jaromela[1], Lato Pezo[2], Biljana Kiprovski[1] and Simona Jaćimović[1]

[1]Institute of Field and Vegetable Crops, National Institute of the Republic of Serbia, Novi Sad, Serbia

[2]Institute of General and Physical Chemistry, University of Belgrade, Belgrade, Serbia

Abstract

Coriander (*Coriandrum sativum* L.) is a multipurpose herb, vegetable, spice and medicinal plant. Its leaves, also called cilantro, are a rich source of vitamins and minerals. They have a specific aroma which resembles bedbugs; nevertheless, they are mostly added to chutneys, salads, and soups or used for preparation of a fresh juice, etc. Ripe seeds have significantly different taste compared to leaves, with citrus overtones, due to essential oils rich in linalool. Important constituents of seeds are proteins and lipids (especially petroselinic acid, monounsaturated omega-12 fatty acid). Seeds could be used as a condiment in pickle spices, seasonings, curry powders, sausages, cakes, pastries, biscuits and buns, beverages, etc. Both leaves and seeds accumulate secondary metabolites, terpenes and phenolics, which possess many pharmacological activities, such as: antioxidant, antimicrobial, anticancer, anti-hyperglycemic, anti-hyperlipidemic, anti-spasmodic and act as a hepatoprotectant. This indicates that coriander can be used as a nutraceutical, a food and a potential remedy for various health issues.

---

[*] Corresponding Author's Email: acimovicbabicmilica@gmail.com.

In: *Coriandrum sativum*: Origin, Uses and Nutrition
Editors: Mamta Pujari, Bharat Kapoor, Neeta Pande et al.
ISBN: 978-1-68507-842-3

Keywords: coriander, essential oil, fatty acids, phenolics, proteins, tocopherols

1. Introduction

Medicinal, aromatic plants and spices (MAPS) are an important source of bioactive compounds, due to which they are widely used in the production of functional food. Functional food, in addition to good nutritional properties, also has beneficial effects on human health, which is of great importance in prevention of various diseases in the modern population.

Coriander (*Coriandrum sativum* L.) belongs to *Apiaceae* family, which has more than 3700 species distributed throughout the world. Because of their characteristic odor, plants from this family are used as a spice and a valuable source of phytochemicals with both nutritional and medicinal properties (Acimovic et al., 2015a; Acimovic et al., 2021a).

Coriander is an annual herbaceous plant with a tap root system and a straight hollow stem, whose basal part has up to 2 cm in diameter. In the flowering period, the height of the stem ranges from 20 to 130 cm. This is a plant with a pronounced heterophylly (Figure 1).

Figure 1. Coriander leaves: (A) the first real leaf; (B) a leaf from rosette; (C) a leaf on the flowering steam; (D) an apical leaf.

Coriander flowers, with white or pale pink petals, are also heterophyllus and organized in an umbel composed of a number of smaller umbels (Figure 2). The central flowers are roundish with small inflexed petals, whereas the peripheral flowers are asymmetric. Petals oriented towards the outside of umbels are lengthened (Figure 2).

Figure 2. Coriander umbels: (A) a central flower; (B) a peripheral flower.

The fruit is a globular or oval cremocarp (a type of schizocarp formed from two one-seeded carpels or mericarps), up to 6 mm in size, brown or yellowish in color (Figures 3 and 4). Most often, cremocarp does not separate spontaneously. Two mericarps have a sclerified pericarp, only on the outside. There is a small carpophore in the central part of the concave fruit. Each mericarp has 6 longitudinal, straight, secondary ribs, which alternate with 5 wavy, barely visible, main ribs. There are only two ventral canals with essential oil in each mericarp (Diederichsen, 1996; Acimovic, 2013; Acimovic, 2014).

The vegetation period of coriander is 104-123 days and it is primarily dependent on the weather conditions (Acimovic, 2013). Vegetative phases: germination (14-19 days), forming of a leaf rosette (30-40 days) and stem elongation (15-20 days) always take more than 50% of vegetation period. The flowering period is 14-25 days, while the ripening period lasts 18-33 days.

Coriander is produced in various agroecological conditions, mainly at smaller production areas. The most important producers of coriander in the world are: India, Mexico, Iran, Morocco, Canada, Romania, Russia and Ukraine (Lal et al., 2014; FAOSTAT, 1994-2019).

Figure 3. Coriander fruit – cremocarp.

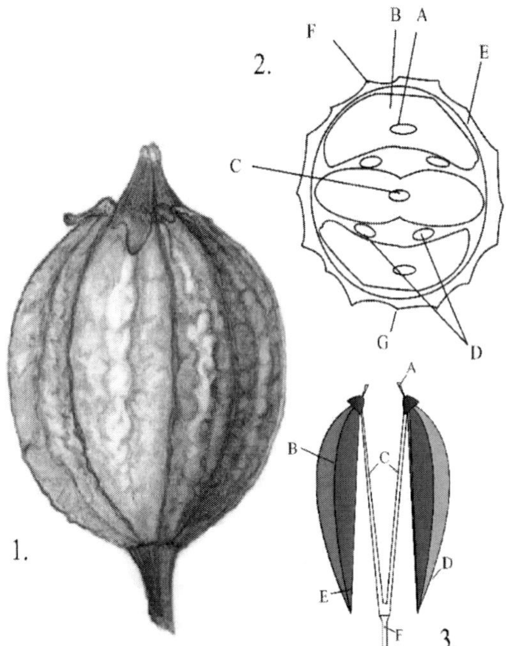

Figure 4. Schemes of coriander fruit: 1. A whole fruit; 2. Horizontal cross section (A) embryo, (B) endosperm, (C) carpophore, (D) vittae, (E) pericarp, (F) right ribs, (G) main ribs; 3. Vertical view of cremocarp (A) stylopod, (B) mericarp, (C) carpophore, (D) dorsal side, (E) commissural side, (F) pedicel.

2. Chemical Composition of Coriander

Coriander is rich in bioactive compounds which are present in the entire plant. These constituents are usually isolated from ripe fruits (*Coriandri fructus*) or the above ground parts of the plant with inflorescences (*Coriandri herba*). Due to a specific organization of ripe fruits (Figure 3), there is an interchangeable use of words fruit and seeds in the literature, when chemical composition is concerned. *C. sativum* fruits have high nutritional value because of lipids (fatty oil), proteins, carbohydrates, fibers and wide range of minerals, trace elements and vitamins (Maurya et al., 2021). However, the quality of coriander fruits and leaves significantly depends on genetic structure, weather conditions during the vegetation period, the content of macro- and micronutrients in the soil, growing conditions and applied agrotechnical measures (Telci et al., 2006, Acimovic et al., 2016a; Krivda et al., 2020).

2.1. Chemical Composition of Coriander Leaves

Besides proteins (1-2% of fresh leaves and 4.05-21.9% dry of leaves), leaves contain lipids (0.4-0.5% of fresh leaves and 4.78-5% of dry leaves), vitamins (A, B6, D) and minerals (K, Mn, Zn, Ca and Mg) (Garg et al., 2004; Shahwar et al., 2012; Ahmadi and Souri, 2018; USDA, 2018). Coriander above ground parts of the plant with inflorescences are used as vegetables.

2.1.1. Leaf Phenolic Compounds

In the blooming stage of coriander, various phenolic compounds were detected in ethanolic extracts of the aboveground part of the plant (leaves and inflorescences): phenolic acids derivatives (ferulic acid, gallic acid, caffeic acid, salicylic acid), flavan-3-ol (catechin), flavonoids (hyperoside, rutin, dihydroquercetin, vicenin, diosmin, hesperidin, luteolin derivatives, apigenin) and coumarins (esculin, esculetin, scopoletin, 4-hydroxycoumarin-umbelliferone) (Oganesyan et al., 2007; Kaiser et al., 2013).

The content of total phenolic compounds is 30 mg/g of dry weight of coriander leaves and 40 mg/g of dry weight of the aboveground parts of plant (leaves and inflorescences). Total flavonoids content in coriander herb (leaves and inflorescences) is 1.8 mg/g of fresh weight. Inflorescence pruning during development of these plants does not affect the content of phenolic compounds and antioxidant activity of coriander extracts (Shahwar et al., 2012; Campos et al., 2019).

2.1.2. Leaf Essential Oil

Coriander leaves contain up to fivefold less essential oil (0.13 to 0.56%) in comparison to coriander fruits and the composition of this essential oil is quite different from that obtained from fruits and seeds (Table 1 and 3). The content of essential oil is greatly determined by growing conditions; for example, foliar application of silicon, biofertilizers and mineral N enhance the content of essential oil and fatty acids content in coriander leaves (Rahimi et al., 2009; Amiripour et al., 2021).

Table 1. The main compounds from coriander leaves essential oil according to reference sources (% of essential oil)

No.	Reference	decanal	1-decanol	E-2-decenal	E-2-dodecenol	tetradecenal	2-dodecenal	cyclodecenol	dodecenal	cyclododecanol	undecanal
1	Shahwar et al. (2012)	1.73	2.18	32.23	7.51	7.57	5.45	0.00	4.07	0.00	2.43
2	Nurzynka-Wierdak (2013)	15.30	4.20	0.60	17.80	2.20	0.00	0.00	4.70	0.00	2.20
3	Nurzynka-Wierdak (2013)	17.20	3.00	0.00	16.50	1.90	0.00	0.00	4.60	0.00	2.10
4	Yildiz (2016)	2.85	3.97	29.87	0.00	2.35	0.00	0.00	5.78	0.00	3.78
5	Turkmen et al. (2016)	11.26	8.29	11.86	1.69	6.64	12.65	8.71	4.50	7.26	3.37
6	Turkmen et al. (2016)	10.81	10.39	10.02	2.20	6.12	10.56	10.24	4.11	7.86	3.73
7	Turkmen et al. (2016)	16.53	13.21	3.48	2.81	14.13	10.40	4.46	7.55	5.14	3.70
8	Turkmen et al. (2016)	9.95	16.16	4.61	5.39	10.11	6.46	8.64	5.67	5.73	2.80
9	Turkmen et al. (2016)	11.94	13.16	10.30	2.06	4.87	7.03	13.78	4.56	5.35	3.77
10	Foudah et al. (2021)	11.04	17.85	6.86	10.45	1.08	1.94	0.00	4.76	0.00	1.40

The most abundant compounds in coriander leaves essential oil are aliphatic aldehydes, which give the essential oil the specific bedbug odor. These compounds are decenal (19.09%) and trans-2-decenal (17.54%). Additionally, leaves contain linalool (13.97%) and α-pinene (1.9%), as well as other derivatives of decenal compound (Shahwar et al., 2012). Published results of the main compounds from coriander leaves essential oil is presented in Table 1.

According to the cluster analysis (unrooted cluster tree) with 10 samples (Table 1) of *Coriandrum sativum L.* essential oil from the literature, it could be concluded that there are several segregated chemotypes of coriander: those with high concentration of E-2-decenal (accessions 1,4); decenal and E-2-dodecenol (accessions 2,3); and those with higher concentrations of 1-decanol (7-10) out of which some have higher than average contents of tetradecenal (7,8), cyclodecenol (9) (Figure 5). Also, there are chemotypes with the mixture of all presented compounds (accessions 5,6) (Table 1).

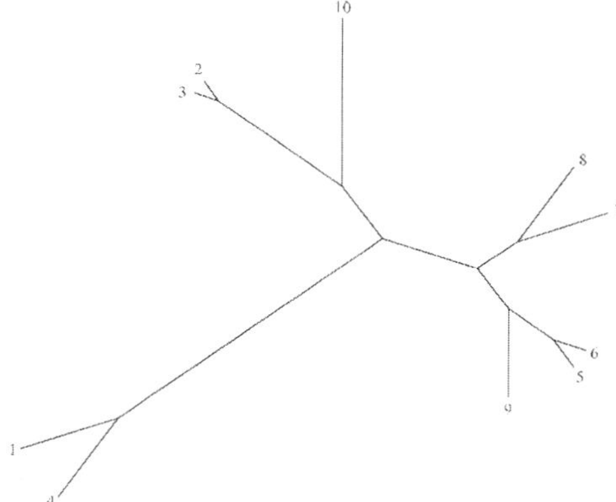

Figure 5. Unrooted cluster tree for different *Coriandrum sativum L.* leaves chemotypes according to reference sources (Table 1).

2.2. Chemical Composition of Coriander Fruits and Seeds

According to various sources (Ramadan and Mörsel, 2002; Shahwar et al. 2012; Sriti et al., 2013; Lal et al., 2014; Jaćimović et al., 2021; USDA, 2018) coriander seeds have following amounts of nutritive compounds: 12.6-17.5% of proteins, 9.12-29% of lipids, 37.1-41.9% of fibers and 6.02-8.59% of ash. Unpublished data of the authors indicate that, in the ten-year monitoring of the quality of coriander collection of the Institute of Field and Vegetable Crops (National Institute of the Republic of Serbia, Novi Sad), the protein content in the seed ranged from 14 to 20% of dry matter.

2.2.1. Seed Fatty Oil

Coriander fatty oil is rich in unsaturated fatty acids, primarily oleic and linoleic (Radusheva et al., 2019), which are two main unsaturated fatty acids that represent over 90% of the total fatty acid content (Jaćimović et al., 2021) (Table 3).

Oleic acid in form of an isomer (18:1, *cis*-Δ6), petroselinic acid, is the dominant fatty acid (75-81.96% of oil) in coriander seeds (Ramadan and Mörsel, 2002; Beyzi et al., 2017; Ashraf et al., 2020; Jaćimović et al., 2021). The fatty acid composition is one of the main indicators of oxidative stability of oil during storage and thermal processing of oil (Radusheva et al., 2019). Oil rich in Polyunsaturated Fatty Acids (PUFA) is prone to oxidation, resulting in instability and short shelf life. In contrast, the high content of stable, monounsaturated (MUFA, Monounsaturated Fatty Acids), oleic acid prolongs the shelf life of the oil, making the oil rich in this fatty acid desirable in the food industry.

Other identified fatty acids in coriander seeds (S1-S4) of the Institute of Field and Vegetable Crops collection are presented in Table 2.

Table 2. Fatty acid composition of analyzed coriander samples (S1-S4) (Jaćimović et al., 2021)

Fatty acid (%)	Number (IUPAC)	S1	S2	S3	S4
Myristic	14:0	0.11	0.10	0.10	0.11
Palmitic	16:0	3.49	3.10	3.44	3.70
Stearic	18:0	1.19	1.00	1.22	1.33
Oleic	18:1(9)	74.54	76.40	75.99	75.25
Linoleic	18:2(9.12)	19.42	18.40	18.25	18.68
Linolenic	18:3(9.12.15)	0.28	0.26	0.25	0.27
Arachidic	20:4(5.8.11.14)	0.19	0.11	0.12	0.14
Eicosapentaenoic	20:5(5.8.11.14.17)	0.38	0.32	0.33	0.29
Behenic	22:0	0.12	0.07	0.07	0.08
Lignoceric	24:0	0.10	0.08	0.11	0.09

2.2.2. Seed Phenolic Compounds

Phenolic compounds play an important role in biological activity and their presence contributes to the antioxidant activity of the plant. Also, these compounds show pronounced antioxidant and antiradical activity in the body of consumers (Rice-Evans et al., 1996). Four isocoumarins were isolated from the *C. sativum* L. full green fruits: coriandrone B, coriandrin, dihydro-coriandrin and coriandrone A (Chen et al., 2009). While coriander mature seeds have vanillin, chlorogenic acid derivatives, catechin derivatives (epicatechin, epicatechin gallate, gallocatechin, epigallocatechin) and flavonoid quercetin-3-O-rutinoside or rutin (Mechchate et al., 2021).

Table 3. The main compounds from coriander seed essential oil according to reference sources (% of essential oil)

No	Reference	linalool	γ-terpinene	α-pinene	camphor	geranyl acetate	geraniol	limonene	β-pinene	camphen	myrcene
1	Bhuiyan et al. (2009)	37.70	14.40	0.00	0.00	17.60	1.90	0.40	1.80	0.10	0.60
2	Zeković et al. (2011)	52.40	7.20	1.70	7.10	4.60	9.00	8.60	0.00	0.30	2.00
3	Shahwar et al. (2012)	55.49	7.47	7.10	5.59	4.24	2.23	4.46	0.71	1.78	0.98
4	Khani and Rahdari (2012)	57.57	0.11	0.12	3.02	15.90	0.24	0.62	0.14	0.00	0.00
5	Hadian et al. (2012)	79.40	6.70	6.70	tr	0.00	0.00	0.10	0.40	tr	0.10
6	Hadian et al. (2012)	74.10	1.20	2.10	0.00	0.00	0.00	0.10	0.30	0.00	0.10
7	Hadian et al. (2012)	89.50	4.00	2.50	0.10	0.00	0.00	0.10	0.20	tr	0.10
8	Hadian et al. (2012)	72.30	8.90	5.60	0.00	0.00	0.00	0.30	0.60	0.00	0.20
9	Hadian et al. (2012)	82.50	5.70	5.00	tr	0.00	0.00	0.10	0.30	tr	0.10
10	Hadian et al. (2012)	79.70	7.00	3.30	0.00	0.00	0.00	0.10	0.00	0.00	0.20
11	Hadian et al. (2012)	69.70	5.30	4.50	0.10	0.00	0.00	0.20	0.50	0.00	0.20
12	Hadian et al. (2012)	78.60	5.10	3.30	0.10	0.00	0.00	0.10	0.40	0.00	0.20
13	Hadian et al. (2012)	85.90	3.10	2.90	0.00	0.00	0.00	0.00	0.00	0.00	0.10
14	Hadian et al. (2012)	78.50	6.40	0.10	0.00	0.00	0.00	0.00	0.00	0.00	0.10
15	Hadian et al. (2012)	83.60	5.00	5.40	0.10	0.00	0.00	0.10	0.50	0.10	0.00
16	Hadian et al. (2012)	88.90	3.20	2.30	0.10	0.00	0.00	0.10	0.20	tr	0.10
17	Hadian et al. (2012)	81.10	7.10	6.00	0.30	0.00	0.00	0.30	0.40	0.10	0.20
18	Hadian et al. (2012)	77.20	10.20	3.30	0.30	0.00	0.00	0.20	0.20	tr	0.10
19	Hadian et al. (2012)	88.10	0.10	0.00	tr	0.00	0.00	0.00	0.00	0.00	0.00
20	Hadian et al. (2012)	85.70	2.60	3.30	0.10	0.00	0.00	0.00	0.30	0.00	0.00
21	Hadian et al. (2012)	80.40	4.10	5.80	0.00	0.00	0.00	0.10	0.40	tr	0.10
22	Hadian et al. (2012)	84.40	3.60	7.40	0.10	0.00	0.00	0.10	0.40	0.00	0.10
23	Hadian et al. (2012)	84.40	4.30	5.10	0.30	0.00	0.00	0.20	0.30	0.00	0.20

Table 3. (Continued)

No	Reference	linalool	γ-terpinene	α-pinene	camphor	geranyl acetate	geraniol	limonene	β-pinene	camphen	myrcene
24	Hadian et al. (2012)	75.60	7.60	7.80	tr	0.00	0.00	0.10	0.50	0.00	0.20
25	Hadian et al. (2012)	71.20	13.20	5.30	0.30	0.00	0.00	0.20	0.20	tr	0.10
26	Hadian et al. (2012)	73.70	5.30	4.50	0.10	0.00	0.00	0.20	0.50	0.00	0.20
27	Hadian et al. (2012)	81.90	6.10	2.30	0.10	0.00	0.00	0.10	0.20	0.00	0.10
28	Hadian et al. (2012)	81.60	7.00	5.40	0.10	0.00	0.00	0.10	0.50	0.10	0.00
29	Hadian et al. (2012)	74.70	5.30	4.50	0.10	0.00	0.00	0.20	0.50	0.00	0.20
30	Hadian et al. (2012)	79.80	8.00	4.50	0.10	0.00	0.00	0.10	0.20	tr	0.10
31	Acimovic et al. (2016a)	72.00	8.60	6.70	3.60	1.10	2.20	1.80	0.60	0.70	0.80
32	Acimovic et al. (2016a)	70.50	6.00	8.20	4.00	1.50	2.50	2.20	0.60	1.00	0.90
33	Acimovic et al. (2016a)	72.00	8.50	6.70	3.60	1.10	2.20	1.80	0.60	0.70	0.80
34	Acimovic et al. (2016a)	69.30	9.60	8.20	3.60	1.10	2.10	1.90	0.70	0.80	0.80
35	Acimovic et al. (2016a)	70.10	9.10	6.80	4.10	1.30	2.30	2.10	0.60	0.80	0.80
36	Acimovic et al. (2016a)	70.90	9.50	7.30	3.30	0.90	2.50	1.70	0.60	0.60	0.70
37	Santos et al. (2019)	64.40	0.00	0.00	2.90	2.10	5.10	0.00	0.00	0.00	0.00
38	Jelen et al. (2021)	44.98	9.47	12.79	6.26	2.68	2.13	0.00	3.01	2.32	0.00

Coriander seeds grown in Serbia had total phenolics content from 3.00 to 3.66 mg gallic acid equivalents (GAE)/g of dry weight (Jaćimović et al., 2021). According to the literature data, the content of phenolics in coriander seeds grown in Turkey (Demir and Korukluoglu, 2020) and Yemen (Al-Mamari, 2002) was 4 and 7 mg GAE/g of dry weight, respectively. Yet, lower contents of total phenolics were recorded in coriander seeds grown in Egypt (0.94 mg GAE/g), Tunisia (1.00 mg GAE/g) and Syria (1.09 mg GAE/g) (Msaada et al., 2017). According to Sriti et al., (2012), total phenolics content in seed, whole fruit and pericarp were 15.55, 12.10 and 2.92 GAE/g, respectively. These variations in phenolics content may be related to genetic variations of the plant, growing conditions, as well as the extraction process of these compounds, since lower contents of total phenolic compounds were recorded in ethanolic, when compared with methanolic extracts of coriander seeds (Demir and Korukluoglu, 2020).

Tocopherols, associated with the prevention of cancer and cardiovascular disease in humans, have high synergistic antioxidant capacity (Shahidi and De Camargo, 2016). The tocopherol content ranges from 14 to 300 mg/L of fatty oil (Sriti et al., 2009; Sahib et al., 2012; Laribi et al., 2015; Jaćimović et al., 2021).

The major sterol compound in fruits, seeds and pericarp is β-sitosterol (36.79, 24.81 and 49.44%, respectively). Additionally, coriander seeds have other sterols, such as: stigmasterol, campesterols, avenasterol, etc. (Sriti et al., 2010; Wei et al., 2019).

2.2.3. Seed Essential Oil

Although the fruits of coriander contain small amount of essential oil (0.1 to 2.9%), it is still the primary product of this crop. The quality and quantity of coriander essential oil may vary depending on a number of factors. The content of essential oil in fruits is greatly determined by genotype or weather conditions, especially drought, due to which the amount of essential oil may diminish considerably (Diederichsen, 1996; Javid et al., 2009; Acimovic et al., 2016a). Also, the significant variations (0.68 to 1.2%) in quantity of essential oil depends on locality, encompassed micro-climatic conditions during periods of flowering and ripening, as well as different types of soil (Acimovic, 2013).

The main compounds of coriander essential oil are: linalool (58.0–80.3%), γ-terpinene (0.3%–11.2%), α-pinene (0.2%–10.9%), p-cymene (0.1%–8.1%), camphor (3.0%–5.1%) and geranyl acetate (0.2%–5.4%) (Raal et al., 2004;

Shahwar et al., 2012, Acimovic, 2014). The main compounds of coriander seed essential oil are shown in Table 3.

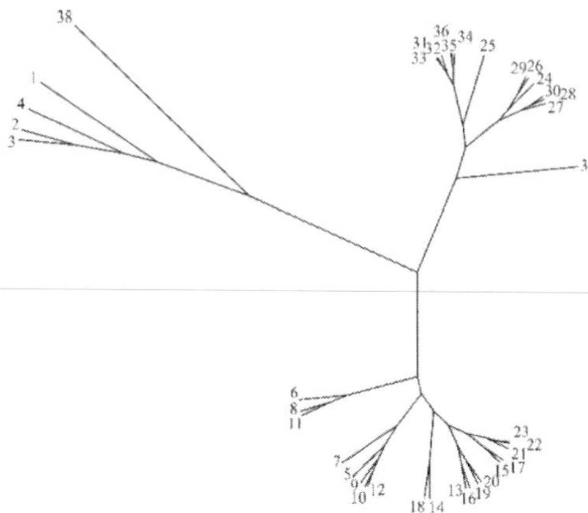

Figure 6. Unrooted cluster tree for different *Coriandrum sativum L* seed chemotypes according to reference sources (Table 3).

According to the cluster analysis (unrooted cluster tree) with 38 samples of main compounds (Table 3) from coriander seed essential oil according to reference sources from cited literature and average values from the authors original results (Figure 6), it could be concluded that there are three segregated groups: (1) those with linalool and geranyl acetate as the major compounds (accessions 1,2,3,4); (2) chemotype with linalool, γ-terpinene and α-pinene in the highest concentration [accessions additionally divided into subgroups according to concentrations of these compounds (5,7,9,10,12); (6,8,11), (13-23)]; and (3) chemotype with linalool, γ-terpinene, α-pinene and camphor in the highest concentration (accessions 24-36). Accession 37 contains linalool and geraniol, while accession 38 contains mixture of linalool, γ-terpinene, α-pinene, camphor and β-pinene. This points to a great diversity of results published on coriander essential oil composition, and general variation in distribution of these compounds.

2.3. Post-Distillation Waste Material

After essential oil distillation, the majority of the plant material remain unused (more than 99%). In some cases, post-distillation material is used as a feed for ruminants, especially milk cows.

Determination of the total phenolics in methanolic extracts (53.78 mg GAE/g of dry extract, on average), which remains after the distillation of essential oil from coriander seeds (post-distillation waste material), and antioxidant activity (0.03 and 0.04 mM TE/g coriander seeds dry extract) of these samples (by the 2,2-diphenyl-1-picrylhydrazil or DPPH radical scavenging method) were presented in Table 4 and represent previously published results of the first author of the chapter (Acimovic et al., 2016b). When compared to other published results commented in the 2.2.2. section of this chapter, it can be concluded that these post-distillation waste of coriander seeds could be of great value to the postharvest processing of coriander seeds.

Table 4. Total phenolics content (TPC) in mg Gallic Acid Equivalents (GAE) and antioxidant capacity (AOC) in mM Trolox Equivalents (TE) of g coriander seeds dry extracts. Six different coriander post-distillation waste material (Acimovic et al., 2016b)

Sample	TPC (mg GAE/g sample)	AOC (mM TE/g sample)
Coriander 1 (local ecotype)	55.7	0.04
Coriander 2 ("dr Josif Pančić" Serbia)	55.9	0.03
Coriander 3 ("Semenarna coop" Serbia)	40.9	0.03
Coriander 4 ("Master seeds" Serbia)	57.2	0.04
Coriander 5 ("Sgaravatti" Italy)	56.2	0.04
Coriander 6 ("Blumen" Italy)	56.8	0.04
Average	53.78	0.037

As for individual phenolic compounds, LC/DAD/MS analysis, confirmed 43 phenolic compounds in the post-distillation waste material of coriander seed (Parejo et al., 2004; Babovic et al., 2010; Hossain et al., 2010; Kaiser et al., 2013; Vallverdu-Quetalz et al., 2014 Acimovic et al., 2016b). Detected compounds were hydroxybenzoic and hydroxycinnamc acids, as well as flavanones and flavanols derivatives.

3. Nutraceutical Potential

A group of food products that has both, nutritional and medicinal properties is most frequently determined as nutraceuticals. It is used as a synonym to functional food; however, the latter usually refers to naturally occurring bioactive molecules in food that have health-promoting properties, and the first includes different food products, dietary supplements and fortified food that is used in disease treatments, additionally to nutritional value.

Bioactive compounds present in coriander are used in traditional and modern medicine, as well as in everyday nutrition. Coriander leaves are used for preparation of chutneys and sauces, green salsas, dips, snacks, soups, while the seeds are used in couscous, stews and salads, as a condiment in pickle spices, seasonings, curry powders, sausages, cakes, pastries, biscuits and buns. Coriander seeds essential oil is used in beverages, baked goods, condiments, relishes and meat products (Acimovic, 2019).

Various studies indicate that coriander enables development of a novel and broad spectrum of antimicrobial herbal formulations for human, animal and food application. Coriander seeds are used as an aromatic, carminative, stomachic, antispasmodic, and against gastrointestinal complaints, such as dyspepsia, flatulence and gastralgia. Coriander is effective against hyperlipidemia and hyperglycemia. It also acts as a hepatoprotectant, and is used in heavy metal detoxification. Apart from this, coriander can be used as an anthelmintic. Its use is recommended for healing of the urinary system, i.e., urethritis, cystitis and urinary tract infections, as well as for insomnia and anxiety. Additionally, it is used as an analgetic and antirheumatic, and as an anticancer agent (Acimovic, 2019).

Due to a specific chemical composition of coriander and its potential, it is recognized as raw material in various branches of industry, such as: pharmaceutical, cosmetics, food, tobacco and beverage industry, production of biodiesel (Illes et al., 2000; Acimovic et al. 2012; Beyzi et al., 2017; Kassahun, 2020).

3.1. Antimicrobial Activity

Natural compounds that can be used as antimicrobial agents became very popular in the past years as alternative medicines, under hypothesis that they have not the risk of side effects (Hamudeng and Serliawat, 2019).

Coriander essential oil demonstrated good anti-bacterial activity (Sambasivaraju and Fazeel, 2016), especially against *Streptococcus pyogenes*, *Staphylococcus aureus* and methicillin resistant *S. aureus*, combined with excellent skin tolerance, while methanol extracts of coriander seeds show good antimicrobial potential against pathogenic bacteria, such as: *Escherichia coli*, *Pseudomonas aeruginosa*, *Staphylococcus aureus*, and *Bacillus pumilus*. Since it effectively kills pathogenic bacteria that cause foodborne diseases and hospital infections (Silva et al., 2011 and 2020), it could be used in antibacterial formulations; as an antiseptic for prevention and treatment of skin infections with Gram-positive bacteria (Casetti et al., 2012; Dua et al., 2014).

3.2. Gastrointestinal System Complaints Treatment

Coriander is generally used for gastrointestinal complaints. Investigations show that coriander fruit crude extract exhibits a dose-dependent spasmogenic effect in guinea-pig ileum and relaxation in rabbit jejunum (Jabeen et al., 2009). Apart from this, the aqueous extracts of coriander leaves exhibit antidiarrheal effect, and could be used for treating diarrhea (Nithya, 2015).

3.3. Hyperlipidemic and Hyperglycemic Potential

Coriander seed powder decreases the uptake and enhances the breakdown of lipids. It can be assumed that coriander has potential to be popularized as a household herbal remedy with preventive and curative effect against hyperlipidemia (Lal et al., 2004). In addition, coriander exhibits high hypolipidaemic activity and provides protection against oxidative stress, and reduces cholesterol deposition in the aorta of animals fed a high cholesterol diet (Joshi et al., 2012). Furthermore, coriander increased the omega-3/omega-6 ratio in the visceral adipose tissue, which may explain its health benefits (Nyakudya et al., 2014).

Coriander seed supplementation was able to improve glycemic indices, lipid profile and oxidative stress status in patients with type 2 diabetes mellitus, and it may be a useful complementary treatment for these patients (Zamany et al., 2021). In addition, the extracts of coriander leaves could protect liver function as it exhibits hypoglycemic, hypolipidemic, and antioxidant effects in diabetic rats, which supports the use of this extracts in managing diabetes mellitus (Sreelatha and Inbavalli, 2012).

3.4. Hepatoprotective Activity

Coriander possesses hepatoprotective potential and can be used for medicinal applications to patients who suffer from liver damage (Acimovic and Milic, 2017). Coriander extracts possess significant anti-fibrotic and antioxidant activity in liver exposed to experimental hepatotoxicity and nephrotoxicity *in vivo*. Increasing coriander consumption is recommended, especially in cases of heavy contaminants and hepatic and renal disorder (Zein et al., 2014). In addition, coriander leaves extracts can reduce the degree of microscopic damage of Wistar rat liver due to mercury exposure (Ardiani et al., 2020).

Ethanolic extracts of coriander fresh leaves possess hepatoprotective activity against carbon tetrachloride induced liver injury. The hepatoprotective effect may be due to its antioxidant potential (Pandey et al., 2011). The ability of the coriander seed water extract to protect the liver against changes mediated by CCl_4 and acetaminophen confirmed that this plant possesses anti-hepatotoxic properties against CCl_4 and acetaminophen induced liver damage in ICR mice (Hewawasam et al., 2016).

3.5. Anthelmintic Properties

Along with many other medicinal properties, coriander seeds essential oil and its main compound, linalool, induced an inhibition of the motility on the third-stage larvae of *Haemonchus contortus*, *Trichostrongylus axei*, *Teladorsagia circumcincta*, *Trichostrongylus colubriformis*, *Trichostrongylus vitrinus* and *Cooperia oncophora* (Helal et al., 2020), which confirmed their anthelmintic activities.

3.6. Urinary System Complaints Treatment

Coriander leaves are used in the traditional medicine as diuretic. Coriander fruits crude extract decreased arterial blood pressure of anesthetized animals, and diuretic activity added value to its use in hypertension prevention (Jabeen et al., 2009). Furthermore, methanolic extract of coriander leaves exhibited significant diuretic activity, corroborated by increased total urine volume and the urine concentration of Na^+, K^+ and Cl^- (Thuraisingam et al., 2019).

3.7. Sedative and Antirheumatic Activities

Coriander seeds has been recommended in traditional medicine for relieving stress and other neurological diseases, as well as for treating insomnia (Alshahrani, 2016). Investigations show that coriander seeds aqueous and alcoholic extracts or essential oil possess sedative-hypnotic activity (Emamghoreishi and Heidari-Hamedani, 2006), almost similar to diazepam (Mahendra and Bisht, 2011). Coriander seed essential oil induces sedative effect which may be due to monoterpene linalool. Therefore, this essential oil could be considered a potential therapeutic agent (Gastón et al., 2016).

Hydroalcoholic extract of fresh coriander leaves exhibits anxiolytic effects against alarm substance-induced locomotor and anxiogenic responses similarly to clonazepam. This data is endorsed by the use of this plant in traditional medicine and provides a potentially new pharmacological intervention for anxiety disorders (Zenki et al., 2020). Furthermore, sedative effect of coriander leaves extracts contributed to the hyperactivity of inhibitory neurons in the brain (according to behavioral study and real-time PCR) (Sakurai et al., 2019).

The anxiolytic effect of aqueous extracts of coriander seeds were examined in male albino mice using elevated plus-maze as an animal model of anxiety, while the effects of the extract on analgesic activity were assessed using the Hot plate method. Results showed that coriander has anxiolytic effect and may have potential analgesic effects, if applied in dose of at 200 mg/kg aqueous extract (Pathan et al., 2011). The aqueous and ethanolic extracts of coriander seeds demonstrated significant analgesic (in Acetic-acid induced writhing method) and anti-inflammatory activities (in Carrageenan induced paw oedema model) (Bhat et al., 2014).

Anti-inflammatory activity of dichloromethane coriander seed extracts was examined using hypotonicity induced human red blood cell membrane stabilization method. It was observed that the coriander extracts exhibited membrane stabilization effect by inhibiting hypotonicity induced lysis of erythrocyte membrane (Kothari et al., 2017). The increased membrane stability also supports the anti-inflammatory activity of coriander, which induced a significantly high antibody response thus reflecting its immune-stimulatory activity (Kothalawala et al., 2020).

3.8. Anticancer Properties

Coriander shows potential in preventing oxidative stress-related diseases and would be useful as supplementation in combination with conventional drugs to enhance treatment of diseases such as cancer (Tang et al., 2013; Acimovic et al., 2018). Coriander seeds powder extracts could be used as a potent anticancer natural medicine and could lead to drug formulation for human breast adenocarcinoma cell lines (MCF-7) (Swetha and Krithika, 2018). Experimental animals, with human hepatocellular carcinoma cell line (HepG2) and mouse melanoma cell line (B16F10), fed with diet containing coriander possessed a smaller number of metastatic regions than those fed with control diet. It was suggested that coriander extract might have the ability to suppress cancer cell migration and invasion, indicating that coriander improves cancer prognosis (Huang et al., 2020).

Conclusion

Additionally, to its wide application as a spice and a vegetable, the use of coriander in traditional and modern medicine positions this valuable plant species in the group of nutraceuticals. Novel studies confirm and explain its potential and activities as a remedy for gastrointestinal (hyperlipidemic, hyperglycemic, hepatoprotective) and urinary systems diseases, in the first place, but also as antimicrobial, sedative, antirheumatic, anthelmintic and anticancer agent.

Future research should focus on testing new ways of using coriander for the preparation of value-added food products, with the aim of promoting high quality diets which supports human health and food security.

References

Acimovic, M. (2019). Nutraceutical potential of Apiaceae. In: Mérillon J. M., Ramawat K. G. (eds.): *Bioactive Molecules in Food, Reference Series in Phytochemistry*, Springer International Publishing AG, pp:1311-1341.

Acimovic, M., Grahovac, M., Stankovic, J., Cvetkovic, M. and Masirevic, S. (2016a): Essential oil composition of different coriander (*Coriandrum sativum* L.) accessions and their influence on mycelial growth of *Colletotrichum* ssp. *Acta Scientiarum Polonorum Hortorum Cultus*, 15(4):35-44.

Acimovic, M., Mara, D., Tešević, V., Stanković, J., Cvetković, M., Urošević, M. and Filipović, V. (2016b). Analysis of total polyphenols from postdestillation waste material of different coriander accessions grown in Serbia. *7th International Scientific Symposium* "Agrosym 2016," October 6-9, 2016, Jahorina, Bosnia and Herzegovina, Book of Proceedings, 796-802.

Acimovic, M. and Milic, N. (2017). Perspectives of the Apiaceae hepatoprotective effects – a review. *Natural Product Communications*, 12(2), 309-317.

Acimovic, M., Oljaca, S. and Drazic, S. (2012). Uses of coriander (*Coriandrum sativum* L.). *Lekovite Sirovine Belgrade*, 31(31), 67-82.

Acimovic, M., Rat, M., Tesevic, V. and Dojcinovic, N. (2018): Anticancer Properties of Apiaceae. In: Petropoulos S., Ferreira I., Barros L. (eds): Phytochemicals in Vegetables: A Valuable Source of Bioactive Compounds, *Bentham Science, pp:*235-254.

Acimovic, M. (2013). *Productivity of caraway, anise and coriander in the system of organic agriculture*. Doctoral dissertation, Faculty of Agriculture, Zemun, University of Belgrade. p. 220 (in Serbian).

Acimovic, M. (2014). *Coriander (Coriandrum sativum L.)*. Zadužbina Andrejević, Belgrade, p. 99.

Acimovic, M., Kostadinović, Lj., Popović, S. and Dojčinović N. (2015a). Apiaceae seeds as functional food. *Journal of Agricultural Sciences, Belgrade*, 60(3), 237-246.

Acimovic, M., Stanković, J. and Cvetković, M. (2016a). Effect of weather conditions, location and fertilization on coriander fruit essential oil quality. *Journal of Essential Oil-Bearing Plants*, 19(5), 1208-1215.

Acimovic, M., Tesevic, V., Stankovic Jeremić, J., Cvetkovic, M., Mara, D., Todosijevic M. and Pezo L. (2021a). *Anethum graveolens*: Chemical Composition, Cultivation and Uses. In: Douglas Isiah (ed.), *Apiaceae – Ecology, uses and toxicity*. Nova Science Publishers, New York, p. 67-132. ISBN: 978-1-53619-060-1.

Ahmadi, M. and Souri, M.K. (2018). Growth and mineral content of coriander (*Coriandrum sativum* L.) plants under mild salinity with different salts. *Acta Physiologiae Plantarum*, 40(11), 1-8. https://doi.org/10.1007/s11738-018-2773-x.

Ali Al-Mamary, M. (2002). Antioxidant activity of commonly consumed vegetables in Yemen. *Malaysian Journal of Nutrition,* 8, 179–189.

Alshahrani, A. M. (2016). *In vivo* neurological assessment of sedative hypnotic effect of *Coriandrum sativum* L. seeds in mice. *International Journal of Phytomedicine*, 8(1):113-116.

Amiripour, A., Jahromi, M. G., Soori, M. K. and Mohammadi Torkashvand, A. (2021). Changes in essential oil composition and fatty acid profile of coriander (*Coriandrum sativum* L.) leaves under salinity and foliar-applied silicon. *Industrial Crops and Products*, 168, 113599. https://doi.org/10.1016/j.indcrop.2021.113599.

Ardiani, Z. R., Sudaryanto, Istiadi, H. and Hadi (2020). The effect of coriander leaves extract on the degree of wistar rats liver microscopic damage was given mercuric chloride orally. *Diponegoro Medical Journal*, 9(6), 436-441.

Ashraf, R., Ghufran, S., Akram, S., Mushtaq, M. and Sultana, B. (2020). Cold pressed coriander (*Coriandrum sativum* L.) seed oil. In: Ramadan M. F. (eds), *Cold Pressed Oils Green Technology, Bioactive Compounds, Functionality, and Applications.* Academic Press, p. 345–356.

Babovic, N., Djilas, S., Jadranin, M., Vlasj, V., Ivanovic, J., Petrovic, S. and Zizovic, I. (2010): Supercritical carbon dioxide extraction of antioxidant fraction from selected Lamiaceae herbs and their antioxidant capacity. *Innovative Food Science and Emerging Technologies*, 11:98-107.

Beyzi, E., Karaman, K., Gunes, A. and Beyzi, S. B. (2017). Change in some biochemical and bioactive properties and essential oil composition of coriander seed (*Coriandrum sativum L.*) varieties from Turkey. *Industrial Crops and Products*, 109, 74–78.

Bhat, S. P., Rizvi, W. and Kumar, A. (2014). *Coriandrum sativum* on pain and inflammation. *International Journal of Research in Pharmacy and Chemistry*, 4(4), 939-945.

Bhuiyan, N. I., Begum, J. and Sultana, M. (2009): Chemical composition of leaf and seed essential oil of *Coriandrum sativum* L. from Bangladesh. *Bangladesh Journal of Pharmacology*, 4:150-153.

Campos, R. A., Junior, S. S., Gonçalves, G. G., Neves, L. G., de Gusmão, S. A., Vianello, F. and Lima, G. P. (2019). Changes in bioactive compounds in spiny coriander leaves in response to inflorescence pruning at different growth stages. *Scientia Horticulturae*, 245, 250-257. https://doi.org/10.1016/j.scienta.2018.10.033.

Casetti, F., Bartelke, S., Biehler, K., Augustin, M., Schempp, C. M. and Frank, U. (2012). Antimicrobial activity against bacteria with dermatological relevance and skin tolerance of the essential oil from *Coriandrum sativum* L. fruits. *Phytotherapy Research*, 26(3), 420-424.

Chen, Q., Yao, S., Huang, X., Luo, J., Wang, J. and Kong, L. (2009). Supercritical fluid extraction of *Coriandrum sativum* and subsequent separation of isocoumarins by high-speed counter-current chromatography. *Food Chemistry*, 117(3), 504-508. doi:10.1016/j.foodchem.2009.04.025.

Demir, S. and Korukluoglu, M. (2020). A comparative study about antioxidant activity and phenolic composition of cumin (*Cuminum cyminum* L.) and coriander (*Coriandrum sativum* L.). *Indian Journal of Traditional Knowledge* (IJTK), 19(2), 383-393.

Diederichsen, A. (1996). *Coriander (Coriandrum sativum L.).* Promoting the conservation and use of underutilized and neglected crops. 3. Institute of Plant Genetics and Crop Plant Research, Gatersleben/International Plant Genetic Resources Institute, Rome. ISBN 9290432845.

Dua, A., Garg, G., Kumar, D. and Mahajan, R. (2014). Polyphenolic composition and antimicrobial potential of methanolic coriander (*Coriandrum sativum*) seed extract. *International Journal of Pharmaceutical Sciences and Research*, 26, 2302-2308.

Emamghoreishi, M. and Heidari-Hamedani, G. (2006). Sedative-hypnotic activity of extracts and essential oil of coriander seeds. *Iranian Journal of Medical Sciences*, 31(1):22-27.

FAOSTAT (FAOSTAT, 1994-2019). http://www.fao.org/faostat/en/#data (accessed on 02 July 2021).

Foudah, A. I., Alqarni, M. H., Alam, A., Salkini, M. A., Ahmed, E. O. I. and Yusufoglu H. S. (2021): Evaluation of the composition and in vitro antimicrobial, antioxidant, and anti-inflammatory activities of Cilantro (*Coriandrum sativum* L. leaves) cultivated in Saudi Arabia (Al-Kharj). *Saudi Journal of Biological Science*, 28, 3461-3468.

Garg, V. K., Singh, P. K. and Katiyar, R. S. (2004). Yield, mineral composition and quality of coriander (*Coriandrum sativum*) and fennel (*Foeniculum vulgare*) grown in sodic soil. *Indian Journal of Agricultural Science*, 74, 221-223.

Gastón, M. S., Cid, M. P., Vázquez, A. M., Decarlini, M. F., Demmel, G. I., Rossi, L. I., Aimar, M. L., and Salvatierra, N. A. (2016). Sedative effect of central administration of *Coriandrum sativum* essential oil and its major component linalool in neonatal chicks. *Pharmaceutical Biology*, 54(10):1954-1961.

Hadian, J., Ebrahimi, S. N., Akramian, M. and Mumivand, H. (2012): Variability of the essential oil content and composition of Iranian landraces of coriander (*Coriandrum sativum* L.), cultivated ina a common environment. *Journal of Essential Oil-Bearing Plants*, 15:89-96.

Hamudeng, A. M, Serliawati. (2019). Effectiveness of antibacterial extract of coriander seeds (*Coriandrum sativum* L.) against *Staphylococcus aureus*. *Journal of Dentomaxillofacial Science*, 4(2): 71-74.

Helal, M. A., Abdel-Gawad, A. M., Kandil, O, M., Khalifa, M. M. E., Cave, G. W. V., Morrison, A. A., Bartley, D. J. and Elsheikha, H. M. (2020). Nematocidal effects of a coriander essential oil and five pure principles on the infective larvae of major ovine gastrointestinal nematodes in vitro. *Pathogens*, 9, 740.

Hewawasam, R. P., Jayatilaka, K. A. P. W., Mudduwa, L. K. B. and Pathirana C. (2016). Modulatory potential of *Coriandrum sativum* on experimentally induced hepatic injury in ICR mice: a biochemical and histopathological investigation. *International Journal of Pharmacognosy and Phytochemical Research*, 8(10), 1709-1716.

Hossain, M. B., Rai, D. K., Brunton, N. P., Martin-Diana, A. B. and Barry-Ryan, C. (2010): Characterization of Phenolic composition in Lamiaceae Spices by LC-ESI-MS/MS. *Journal of Agricultural and Food Chemistry*, 58:10576-10581.

Huang, H., Nakamura, T., Yasuzawa, T., and Ueshima, S. (2020). Effects of *Coriandrum sativum* on migration and invasion abilities of cancer cells. *Journal of Nutritional Science and Vitaminology* (Tokyo), 66(5), 468-477.

Illes, V., Daood, H. G., Perneczki, S., Szokonya, L. and Then, M. (2000). Extraction of coriander seed oil by CO_2 and propane at super - and subcritical conditions. *The Journal of Supercritical Fluids,* 17, 177–186.

IUPAC *Lipid nomenclature: Appendix A: names of and symbols for higher fatty acids*. https://www.qmul.ac.uk/sbcs/iupac/lipid/appABC.html.

Jabeen, Q., Bashir, S., Lyoussi, B. and Gilani, A. H. (2009). Coriander fruit exhibits gut modulatory, blood pressure lowering and diuretic activities. *Journal of Ethnopharmacology*, 122, 123-130.

Jaćimović, S., Marjanović Jeromela, A., Kiprovski, B., Zeremski, T., Grahovac, N. and Acimovic, M. (2021). Nutritional quality of coriander from the collection of the Institute of Field and Vegetable Crops. Book of abstracts. 62nd *Congress of production and processing of oilseeds*, 27.06-02.07.2021., Herceg Novi, Montenegro, p. 179-186 (in Serbian).

Javid, H., Khan, A. L., Farmanullah, K., Hussain, S. T., and Shinwari, Z. K. (2009). Proximate and nutrient investigations of selected medicinal plants species of Pakistan. *Pakistan Journal of Nutrition*, 8(5), 620-624.

Jelen, H. H., Marciankowska, M. and Marek, M. (2021): Determination of volatile terpenes in coriander cold pressed oil by vacuum assisted sorbent extraction (VASE). *Molecules*, 26:884.

Joshi, S. C., Sharma, N. and Sharma, P. (2012). Antioxidant and lipid lowering effects of *Coriandrum sativum* in cholesterol fed rabbits. *International Journal of Pharmacy and Pharmaceutical Sciences*, 4(3), 231-234.

Kaiser, A., Kammerer, D. R. and Carle, R. (2013). Impact of blanching on polyphenol stability and antioxidant capacity of innovative coriander (*Coriandrum sativum* L.) pastes. *Food Chemistry*, 140(1-2), 332-339. http://dx.doi.org/10.1016/j.foodchem.2013.02.077.

Kassahun, B. M. (2020). Unleashing the Exploitation of Coriander (*Coriander sativum* L.) for Biological, Industrial and Pharmaceutical Applications. *Academic Research Journal of Agricultural Science and Research,* 552–564.

Khani, A. and Rahdari, T. (2012): Chemical composition and insecticidal activity of essential oil from *Coriandrum sativum* seeds against *Tribolium confusum* and *Callosobruchus maculatus*. *ISRN Pharmaceutics*, 263517.

Kothalawala, S. D., Edward, D., Harasgama, J. C., Ranaweera, L., Weerasena, O. V. D. S. J., Niloofa, R., Ratnasooriya, W. D., Premakumara G. A. S. and Handunnetti, S. M. (2020). Immunomodulatory activity of a traditional Sri Lankan concoction of *Coriandrum sativum* L. and *Coscinium fenestratum* G. *Evidence-Based Complementary and Alternative Medicine*, 2020, 9715060, https://doi.org/10.1155/2020/9715060.

Kothari, S., Priya, V. V. and Gayathri, R. (2017). Anti-inflammatory activity of *Coriandrum sativum* using HRBC membrane stabilizing method. *International Journal of Pharmaceutical Sciences Review and Research*, 43(2), 68-70.

Krivda, S. I., Nevkrytaya, N. V., Pashtetsky, V. S., Babanina, S. S., Skipor, O. B., Krivchik, N. S. and Skiba, A. V. (2020). Analysis of the collection of *Coriandrum sativum* L. as a source of high-potential samples for selection research. *International Journal of Biology and Biomedical Engineering*, 14, 63-69; DOI: 10.46300/91011.2020.14.10.

Lal, A. A., Kumar, T., Murthy, P. B. and Pillai K. S. (2004). Hypolipidemic effect of *Coriandrum sativum* L. in triton-induced hyperlipidemic rats. *Indian Journal of Experimental Biology*, 42, 909-912.

Lal, G., Saran, P. L., Devi, G., Deepak and Raj, R. (2014). Seed production technology of coriander (*Coriandrum sativum*). *Advances in vegetable agronomy* (First Edition). PGS, IARI and DARE, ICAR, New Delhi. p. 214–222.

Laribi, B., Kouki, K., M'Hamdi, M. and Bettaieb, T. (2015). Coriander (*Coriandrum sativum L.*) and its bioactive constituents. *Fitoterapia,* 103, 9–26.

Mahendra, P., and Bisht, S. (2011). Anti-anxiety activity of *Coriandrum sativum* assessed using different experimental anxiety models. *Indian Journal of Pharmacology*, 43(5):574-577.

Maurya, U., Yadav, P., Maurya, S., Maurya, R. and Yadav, P. (2021). *Coriandrum sativum* L.: A nutritional food and phytomedicine. *International Journal of Creative Research Thoughts*, 9(1), 1730-1740.

Mechchate, H., Es-safi, I., Amaghnouje, A., Boukhira, S., Alotaibi, A. A., Al-zharani, M., Nasr, F. A., Noman, O. M., Conte, R., El Hamsas El Youbi, A., Bekkari, H. and Bousta, D. (2021). Antioxidant, anti-inflammatory and antidiabetic proprieties of LC-MS/MS identified polyphenols from coriander seeds. *Molecules,* 26, 487.

Msaada, K., Ben Jemia, M., Salem, N., Bachrouch, O., Sriti, J., Tammar, S., Bettaieb, I., Jabri, I., Kefi, S., Limam, F. and Marzouk, B. (2017). Antioxidant activity of methanolic extracts from three coriander (*Coriandrum sativum* L.) fruit varieties. *Arabian Journal of Chemistry,* 10, 3176–3183.

Nithya, V. (2015). Evaluation of antidiarrheal activity on *Coriandrum sativum* Linn., in Wistar albino rats. *World Journal of Pharmaceutical Research*, 4(5), 638-643.

Nurzynka-Wierdak, R. (2013): Essential oil composition of the coriander (*Coriandrum sativum* L.) herb depending on the development stage. *Acta Agrobotanica*, 66(1), 53-60.

Nyakudya, T., Makaula, S., Mkumla, N. and Erlwanger K. (2014). Dietary supplementation with coriander (*Coriandrum sativum*) seed: effect on growth performance, circulating metabolic substrates and lipid profile of the liver and visceral adipose tissue in healthy female rats. *International Journal of Agriculture and Biology*, 16, 125-131.

Oganesyan, E. T., Nersesyan, Z. M. and Parkhomenko, A. Y. (2007). Chemical composition of the above-ground part of *Coriandrum sativum*. *Pharmaceutical Chemistry Journal*, 41, 149–153. https://doi.org/10.1007/s11094-007-0033-2.

Pandey, A., Bigoniya, P., Raj, V. and Patel, K. K. (2011). Pharmacological screening of *Coriandrum sativum* Linn. for hepatoprotective activity. *Journal of Pharmacy and Bioallied Sciences*, 3(3), 435-441.

Parejo, I., Jauregui, O., Sanchez-Rabaneda, F., Viladomat, F., Bastida, J. and Codina, C. (2004): Separation and charaterization of phenolic compounds in fennel (*Foeniculim vulgare*) using liquid chromatography-negative electrospray ionization tandem mass spectrometry. *Journal of Agricaltural and Food Chemistry*, 52:3679-3687.

Pathan, A. R., Kothawade, K. A. and Logade, M. N. (2011). Anxiolytic and analgesic effect of seeds of *Coriandrum sativum* Linn. *International Journal of Research in Pharmacy and Chemistry*, 1(4), 1087-1099.

Prachayasittikul, V., Prachayasittikul, S., Ruchirawat, S. and Prachayasittikul, V. (2017). Coriander (*Coriandrum sativum*): A promising functional food toward the well-being. *Food Research International,* 105, 305–323.

Raal, A., Arak, E. and Orav, A. (2004). Koriandriviljade eeterliku õli keemiline koostis ja vastavus Euroopa farmakopöa nõuetele [Chemical composition of coriander essential oil and compliance with European Pharmacopoeia requirements]. *Agraarteadus*, 15(4), 234-239.

Radusheva, P., Pashev, A., Uzunova, G., Nikolova, K., Gentscheva, G., Perifanova, M. and Marudova, M. (2019). Physicochemical characteristics of seed oil of *Sambucus ebulus*, *Coriandrum sativum* L. and *Silybum marianum* L. *Bulgarian Chemical Communications,* 51, 144–149.

Rahimi, A. R., Mashayekhi, K., Amini, S. and Soltani, E. (2009). Effect of mineral vs. biofertilizer on the growth, yield and essential oil content of coriander (*Coriandrum sativum* L.). *Medicinal and Aromatic Plant Science and Biotechnology*, 3, 21-23.

Ramadan, M. F. and Mörsel, J. T. (2002). Oil composition of coriander (*Coriandrum sativum* L.) fruit-seeds. *European Food Research and Technology,* 215, 204–209.

Rice-Evans, C. A., Miller, N. J. and Paganga, G. (1996). Structure-antioxidant activity relationships of flavonoids and phenolic acids. *Free Radical Biology and Medicine,* 20, 933–956.

Sahib, N. G., Anwar, F., Gilani, A. H., Hamid, A. A., Saari, N. and Alkharfy, K. M. (2012). Coriander (*Coriandrum sativum* L.): A Potential Source of High - Value Components for Functional Foods and Nutraceuticals - A Review. *Phytotherapy research,* 10, 1439-56.

Sakurai, J., Izumo, N. and Watanabe, Y. (2019). Effect of *Coriandrum sativum* L. leaf extract on the brain GABA neurons in mice. *Journal of Nutritional Health and Food Science*, 7(2):1-7.

Sambasivaraju, D., and *Fazeel, Z. A. (2016).* Evaluation of antibacterial activity of *Coriandrum sativum* (L.) against gram – positive and gram – negative bacteria. *International Journal of Basic and Clinical Pharmacology*, 5(6), 2653-2656.

Santos, M. D., Neto, M. F. C., Melo, A. C. R., Takarashi, J. A., Ferraz, V. P., Chagas, E. A., Chagas, P. C. and Filho, A. A. M. (2019): Chemical composition of essential oil of coriander seed (*Coriandrum sativum*) cultivated in Amazon Savannah, Brazil. *Chemical Engineering Transaction*, 75:409-414.

Shahidi, F. and De Camargo, A. C. (2016). Tocopherols and tocotrienols in common and emerging dietary sources: occurrence, applications, and health benefits. *International Journal of Molecular Sciences,* 17(10), 174.

Shahwar, M. K., El-Ghorab, A. H., Anjum, F. M., Butt, M. S., Hussain, S. and Nadeem, M. (2012): Characterization of coriander (*Coriandrum sativum* L.) seeds and leaves: volatile and nonvolatile extracts. *International Journal of Food properties,* 15:736-747.

Silva, F., Ferreira, S., Queiroz, J. A. and Domingues, F. C. (2011). Coriander (*Coriandrum sativum* L.) essential oil: its antibacterial activity and mode of action evaluated by flow cytometry. *Journal of Medical Microbiology*, 60, 1479-1486.

Silva, F., Domeño, C. and Domingues, F. C. (2020). *Coriandrum sativum* L.: characterization, biological activities, and applications. In: Predi V. & Watson R. (eds.), *Nuts and Seeds in Health and Disease Prevention* (Second Edition). Academic Press, p. 511–514.

Sreelatha, S., and Inbavalli, R. (2012). Antioxidant, antihyperglycemic, and antihyperlipidemic effects of *Coriandrum sativum* leaf and stem in alloxan-induced diabetic rats. *Journal of Food Science*, 77(7), T119-T123.

Sriti, J., Wannes, W. A., Talou, T., Jemia, M. B., Kchouk, M. E. and Marzouk, B. (2012): Antioxidant properties and polyphenol contents of different parts of coriander (*Coriandrum sativum* L.) fruit. *La Rivista Italiana Delle Sostanze Grasse*, 49:253-262.

Sriti, J., Neffati, M., Msaada, K., Talou, T. and Marzouk, B. (2013). Biochemical characterization of coriander cakes obtained by extrusion. *Journal of Chemistry, ID 871631.* https://doi.org/10.1155/2013/871631

Sriti, J., Talou, T., Wannes, W. A., Cerny, M. and Marzouk, B. (2009). Essential oil, fatty acid and sterol composition of Tunisian coriander fruit different parts. *Journal of the Science of Food and Agriculture*, 89(10), 1659–1664. https://doi.org/10.1002/jsfa.3637.

Sriti, J., Wannes, W. A., Talou, T., Mhamdi, B., Hamdaoui, G. and Marzouk, B. (2010). Lipid, fatty acid and tocol distribution of coriander fruit's different parts. *Industrial Crops and Products*, 31(2), 294–300. DOI 10.1007/s11746-009-1505-1.

Swetha, M. and Krithika, N. (2018). *In vitro* cytotoxicity and cell viability assay of *Coriandrum sativum* L. seed powder extracts. *World Journal of Pharmaceutical Research*, 7(7), 317-322.

Tang, E. L. H., Rajarajeswaran, J., Fung, S. Y. and Kanthimathi, M.S. (2013). Antioxidant activity of *Coriandrum sativum* and protection against DNA damage and cancer cell migration. *BMC Complementary and Alternative Medicine*, 13, 347 http://www.biomedcentral.com/1472-6882/13/347.

Telci, I., Toncer, O.G. and Sahbaz, N. (2006). Yield: essential oil content and composition of *Coriandrum sativum* varieties (var. *vulgare* Alef and var. *microcarpum* DC) grown in two different locations. *Journal of Essential Oil Research,* 18, 189–193.

Thuraisingam, S., Sunilson, J. A. J., Kumari, A. V. A. G. and Anandarajagopal, K. (2019). Preliminary phytochemical analysis and diuretic activity of the extracts of *Coriandrum sativum* leaves in Wistar albino rats. *International Research Journal of Pharmacy and Medical Sciences*, 3(1), 1-3.

Turkmen, M., Bahadirli, N. P. and Mert, A. (2016): Essential oil components of fresh coriander (*Coriandrum sativum* L.) herbs from different locations in Turkey. 6th *International Conference on Advanced Materials and Systems*, doi:10.24264/icams-2016.II.19.

USDA (accessed on 02 July 2021):

- *Spices, coriander seed* https://fdc.nal.usda.gov/fdc-app.html#/food-details/170922/nutrients
- *Spices, coriander leaf, dried* https://fdc.nal.usda.gov/fdc-app.html#/food-details/170921/nutrients
- *Coriander (cilantro) leaves, raw* https://fdc.nal.usda.gov/fdc-app.html#/food-details/169997/nutrients

Vallverdu-Quetalz, A., Reguerio, J., Martinez-Huelamo M., Alvarenga, J. F. R., Neto Leal, L. and Lamuela-Raventos, R. M. (2014): A comprehensive study on phenolic profile of widely used culinary herbs and spices: Rosemary, thyme, oregano, cinnamon, cumin and bay. *Food Chemisty*, 154:299-307.

Wei, J. N., Liu, Z. H., Zhao, Y. P., Zhao, L. L., Xue, T. K. and Lan, Q. K. (2019). Phytochemical and bioactive profile of *Coriandrum sativum* L. *Food Chemistry,* 286, 260-267.

Yildiz H. (2016). Chemical composition, antimicrobial, and antioxidant activities of essential oil and ethanol extract of *Coriandrum sativum* L. leaves from Turkey. *International Journal of Food Properties*, 19(7), 1593-1603.

Zamany, S., Mahdavi, A. M., Pirouzpanah, S. and Barzegar, A. (2021). The effects of coriander seed supplementation on serum glycemic indices, lipid prole and parameters of oxidative stress in patients with type 2 diabetes mellitus: a randomized double-blind placebo-controlled clinical trial. *Research Square* https://doi.org/10.21203/rs.3.rs-262149/v1.

Zein, N., Elghani, E. A. and Talat, E. (2014). Effect of *Coriadrum sativum* on experimentally iduced hepatotoxicity of carbon tetrachloride in rats. *Biochemistry Letters*, 9(11), 135-155.

Zeković, Z., Adamović, D., Ćetković, G., Radojković, M. and Vidović, S. (2011): Essential oil and extract of coriander (*Coriandrum sativum* L.). *Acta Periodica Technologica*, 42:281-288.

Zenki, K. C., de Souza, L. S., Góis, A. M., Lima, B. S., Araújo, A. A. S., Vieira, J. S., Camargo, E. A., Kalinine, E., de Oliveira, D. L. and Walker, C. I. B. (2020). *Coriandrum sativum* extract prevents alarm substance-induced fear- and anxiety-like responses in adult zebrafish. *Zebrafish*, 17(2): 120-130.

Chapter 4

Phytochemistry and Pharmacological Activities of *Coriandrum sativum* L.

B. Akshatha, W. N. Sudheer and N. Praveen*

Department of Life Sciences, CHRIST (Deemed to be University), Bengaluru, India

Abstract

Coriandrum sativum L. is a pharmaceutically significant herb that is used for culinary purposes and in herbal formulations. It is an annual herb from the family Apiaceae (Umbelliferae) with unique taxonomic characters. Generally called Dhania or kutumbari, it is cultivated worldwide for its distinct flavors and medicinal properties. Coriander is a rich reservoir of nutrients and significant biochemicals. The phenomenal healing properties of coriander can be attributed to the phytochemicals present in essential oils produced in various parts of the plants, such as leaves, flowers, fruit, and seed. The essential oils are rich in biochemicals like Linalool, (E)-2-decenal, 2-Decenoic acid, camphor, etc. These biomolecules altogether contribute to many pharmacological activities like analgesic, anticancer, anticonvulsant, antidiabetic, anthelmintic, antihypertensive, antimicrobial, anti-mutagenic, antioxidant, anxiolytic, diuretic, hypnotic activity. There are many scientifically proven reports to suggest its importance for usage. The present chapter summarizes the nutrition, biochemicals, and the scientifically proven pharmacological activities of *Coriandrum sativum* L as well as its cultivation and processing.

* Corresponding Author's Email: praveen.n@christuniversity.in.

In: *Coriandrum sativum*
Editors: Mamta Pujari, Bharat Kapoor, Neeta Pande et al.
ISBN: 978-1-68507-842-3

Keywords: coriander, essential oils, biomolecules, cultivation, and pharmacological activities

1. Introduction

Human civilization evolved with the plants and trees around them. There are many herbs, shrubs, and trees which are used by humans for their daily purposes like food, shelter, and clothing. There are several herbs which have been used as food crops for ages for their medicinal value. *Coriandrum sativum* L. (coriander) is one such herb used for cooking and also has many scientific reports for its medicinal value (Laribi et al., 2015).

There are various scientific names for coriander but among them, *Coriandrum sativum* L. is most accepted. Other synonyms are *Coriandrum majus* Gouan, *Coriandrum diversifolium* Gilib., *Coriandrum testiculatum* Lour., *Selinum Coriandrum* (Vest) E.H.L.Krause (Diederichsen and Hammer, 2003). Along with scientific names, there are many common names or vernacular names for coriander. In Sanskrit it is called Kusthumbari, in Hindi, it is dhania, in Tamil, it is Kottamalli, in Telugu it is Kottimera, and in Kannada, it is called Kottambari. Taxonomically *Coriandrum sativum* L. belongs to the family Apiaceae (Umbelliferae) (Diederichsen and Hammer, 2003). Coriander is distributed all over the world and it is grown annually in all seasons. Reports suggest that *Coriandrum sativum* L. originated from northeastern regions and central Asian countries. Initially, this plant was regarded as the weed found in cereal crop fields. Later, domestication of the crop and many varieties were developed from the wild types. The morphology of coriander is unique and it is very easy to identify (Diederichsen, 1996). The table mentioned below shows the detailed characteristics of *Coriandrum sativum* L.

Coriander is mostly used for cooking purposes all over the world. It is used for making pickles, soups, curry, and for seasoning. Along with food-grade uses, there are many reports suggested for their medicinal properties. There are diversified essential oils present in the plant body which have various phyto metabolites with medicinal value. Coriander is high in nutrition and is rich in various types of minerals, vitamins and proteins. It is seen that all parts of the plant body are rich in plant secondary metabolites and account for different medicinal properties like anti-microbial, anti-oxidant, anti-diabetic, anti-convulsant, and hypnotic activity (Onder, 2018). There are many commercially important secondary metabolites present in the plant body which are used in formulating medicines. Coriander is used in the perfume industry for its aroma and useful in the production of lotions and perfumes (Mandal and Mandal, 2015). Food oil extracted from coriander is used as a preservative to reduce the lipid peroxidation of processed meat products. The antifungal property of coriander extracts helps to overcome bacterial contaminations. So, coriander extracts are extensively used in the food industry (Silva et al., 2020). The detailed morphological traits have been tabulated in Table 1.

Table 1. Detailed morphological characteristics of *Coriandrum sativum* L. (Diederichsen, 1996)

Sl. No	Morphological terms	Characteristics
1	Habit	The herbaceous plant grows between 0.20 to 1.40 meters
2	Habitat	Terrestrial habitat and grows in summer and winter
3	Leaf	Leaves are green and light green and they are very thin and are present in different shapes - diversifolius and arranged alternatively. The leaves arranged in the basal region are tri pinnatifid, the leaves arranged above them showed more pinnatifid and the leaves arranged on the apical region are narrow lanceolate in shape.
4	Stem	Green tender stems with sympodial growth in a monochasial branching pattern.
5	Root	Taproot and have numerous root hairs
6	Inflorescence	Compound umbel and umbellets are arranged at the same level.
6	Flower	Peripheral flowers in umbellets germinate followed by others. Petals -5 in number and they are white and pink with an inflexed nature. The flowers in the peripheral region in the umbellate are protandrous and the central ones are sterile or staminate. Five filaments are present and arranged between the petals. Pistils are 2 in number and the position of the ovary is inferior. Pollination happens commonly through the mode of geitonogamy.
7	Fruit	Brown in color and are globular shaped and have a dry fibrous texture. It is a schizocarp type of fruit. The seeds developed in these fruits show epigeal germination.

2. Nutritional and Chemical Composition

Coriander is very rich in nutrition and scientific reports suggest that all the parts of the plant are composed of different vitamins, minerals and have good moisture content. It is observed that vitamins like riboflavin, niacin, ascorbic acid (vitamin C), retinol (Vitamin A) are present in the leaves of coriander. Fresh leaves of coriander are reported to have minerals like calcium (0.14%), phosphorus (0.06%), and iron (0.01%). Along with leaves, seeds also possess good nutritional value. Seeds have crude protein of around 11.49% and 28.43% of crude fibre and 10.53% of starch (Nadeem et al., 2013). Coriander is rich in bioactive molecules, which aid in different medicinal properties. It has been observed that the essential oil content varies in different stages of the phenology i.e., vegetative, full flowering, immature green fruit, and mature fruits. The highest oil yield of 0.37% has been reported in the immature green fruit (Ramezani et al., 2009). These essential oils have been used as the flavoring agent in food materials and physically they are colorless or yellow with peculiar odor and flavor (Burdock and Carabin, 2009).

2.1. Phytochemistry of Coriander

2.1.1. Leaf

Gas chromatography studies suggest that essential oils extracted from the leaves of the coriander using liquid-liquid extraction (hydro-distillation) showed various volatile compounds (Iqbal et al., 2017). Compounds like (E)-2-decenal, linalool, (E)-2-dodecenal, (E)-2- tetradecenal, 2-decen-1-ol, (E)-2-undecenal, dodecanal, (E)-2-tridecenal, (E)-2-hexadecenal, pentadecenal, and α-pinene were reported. Among these (E)-2-decenal, linalool was reported to be present more than 10% by area in the chromatograph. Along with volatile compounds, non-volatile compounds were also analyzed from the leaves and seeds of coriander and it is observed that they are highly soluble in methanol than any other solvent (Shahwar et al., 2012). In another study, GC-MS analysis reveals that there are diversified compounds present in the essential oils extracted from leaves of coriander. Percentage of total ion current suggests that on whole aldehydes were around 82.6% and alcohols account for 16% and it is analysed that (E)-2-decenal covers 46.1% in percentage of total ion current (Potter and Fagerson, 1990).

Different cultivars of coriander from Bangladesh were subjected to biochemical characterization of their essential oils. It is seen that 2-Decenoic acid, E-11- tetradecenoic acid, capric acid were present more than 10% in the yield and it is seen that 2-Decenoic acid was present in high amounts and it was around 30.8% (Bhuiyan et al., 2009). In the aerial parts of coriander, isocoumarins like coriandrone A, B (Baba et al., 1991), C, D, E (Taniguchi et al., 1996) were isolated which show diversified pharmacological properties. Carotenoids like β-carotene, β-cryptoxanthin epoxide, lutein- 5,6-epoxide, violaxanthin, and neoxanthin which show great anti-oxidant were isolated from the ether extracts of coriander (Guerra et al., 2005). Gas chromatography–mass spectroscopy (GC-MS) analysis of essential oils extracted from leaves revealed cyclododecanol, and previously reported decanals (Chung et al., 2012). Polyphenolic compound such as quercetin-glucuronide is abundant, contributing considerably to coriander's pharmacological action (Kaiser et al., 2013). There are other polyphenols like caffeic, ferulic, gallic, and chlorogenic acids being reported in the coriander leaves (Abascal and Yarnell, 2012).The phytochemicals that have been analyzed using gas chromatography have been tabulated in Table 2.

Table 2. Phytochemicals analyzed using gas chromatography

Sl. No.	Phytochemicals analyzed using GC-MS	Organ	Reference
1	α-pinene	Leaf, Seed	Shahwar et al. (2012); Momin et al. (2012)
2	Decanal	Leaf, Seed, Stem	Shahwar et al. (2012); Potter and Fagerson, 1990; Bhuiyan et al. (2009); Chung et al. (2012)
3	(E)-2-decenal	Leaf	Shahwar et al. (2012); Potter and Fagerson (1990)
4	2-decen-1-ol	Leaf	Shahwar et al. (2012); Potter and Fagerson (1990)
5	1-Decanol	Leaf	Shahwar et al. (2012); Potter and Fagerson (1990); Chung et al. (2012)
6	Linalool	Leaf, Seed, Fruit	Shahwar et al. (2012); Nagella et al. (2012); Bhuiyan et al. (2009); Momin et al. (2012)
7	Undecanal	Leaf	Shahwar et al. (2012)
8	(E)-2-undecenal	Leaf	Shahwar et al. (2012); Potter and Fagerson (1990)
9	Dodecanal	Leaf	Shahwar et al. (2012); Potter and Fagerson (1990); Bhuiyan et al. (2009), Chung et al. (2012)
10	1-Eicosanol	Leaf	Shahwar et al. (2012)
11	(E)-2-dodecenal	Leaf	Shahwar et al. (2012); Potter and Fagerson (1990)
12	Tridecenal	Leaf, Stem	Shahwar et al. (2012); Chung et al. (2012)
13	(E)-2-tridecenal	Leaf	Shahwar et al. (2012)
14	Tetra decenal	Leaf, Stem	Shahwar et al. (2012); Chung et al. (2012)
15	(E)-2- tetradecenal	Leaf	Shahwar et al. (2012)
16	Carvone	Leaf	Shahwar et al. (2012)

Table 2. (Continued)

Sl. No.	Phytochemicals analyzed using GC-MS	Organ	Reference
17	Pentadecenal	Leaf	Shahwar et al. (2012)
18	(E)-2-Hexadecenal	Leaf	Shahwar et al. (2012)
19	Camphene	Seed	Shahwar et al. (2012)
20	D-Limonene	Seed	Shahwar et al. (2012)
21	Limonene	Seed	Shahwar et al. (2012)
22	γ-terpinene	Seed	Shahwar et al. (2012); Bhuiyan et al. (2009); Momin et al. (2012)
23	Camphor	Seed	Shahwar et al. (2012); Momin et al. (2012)
24	Geraniol	Seed	Shahwar et al. (2012); Bhuiyan et al. (2009)
25	Geranyl acetate	Seed	Shahwar et al. (2012); Bhuiyan et al. (2009); Momin et al. (2012)
26	8-methyl-2-noneal	Leaf	Potter and Fagerson (1990)
27	5-Isopropenyl-2-methylcyclopent-1-enecarboxaldehyde	Seed	Nagella et al. (2012)
28	Linalool oxide	Seed	Nagella et al. (2012)
29	3,7-Octadiene-2,6-diol, 2,6-dimethyl-	Seed	Nagella et al. (2012)
30	2,6-Octadien-1-ol, 3,7-dimethyl-, acetate, (E)-	Seed	Nagella et al. (2012)
31	Hotrienol	Seed	Nagella et al. (2012)
32	7-Oxabicyclo[4.1.0]heptanes, 1-methyl-4-(2-methyloxiranyl)-	Seed	Nagella et al. (2012)
33	2-Decanoic acid	Leaf	Bhuiyan et al. (2009)
34	2-Dodecenal	Leaf	Bhuiyan et al. (2009); Chung et al. (2012)
35	2-Tridecenal, (E)	Leaf	Bhuiyan et al. (2009)
36	2-Undecenal	Leaf	Bhuiyan et al. (2009)
37	Cyclododecane	Leaf	Bhuiyan et al. (2009)
38	Decamethylene glycol	Leaf	Bhuiyan et al. (2009)
39	Dodecanoic acid	Leaf	Bhuiyan et al. (2009)
40	E-11-tetradecanoic acid	Leaf	Bhuiyan et al. (2009)
41	Capric acid	Leaf	Bhuiyan et al. (2009)
42	Nonanoic acid	Leaf	Bhuiyan et al. (2009)
43	E-Undecanoic acid	Leaf	Bhuiyan et al. (2009)
44	Undecanoic acid	Leaf	Bhuiyan et al. (2009)
45	Undecyl alcohol	Leaf	Bhuiyan et al. (2009)
46	α-Pinene	Seed	Bhuiyan et al. (2009)
47	m-Cymene	Seed	Bhuiyan et al. (2009)
48	Citronellal	Seed	Bhuiyan et al. (2009)
49	Citronellol	Seed	Bhuiyan et al. (2009)
50	Citral	Seed	Bhuiyan et al. (2009)
51	Citronellyl acetate	Seed	Bhuiyan et al. (2009)
52	Curcumene	Seed	Bhuiyan et al. (2009)

Sl. No.	Phytochemicals analyzed using GC-MS	Organ	Reference
53	α Cedrene	Seed	Bhuiyan et al. (2009)
54	α Farnesene	Seed	Bhuiyan et al. (2009)
55	β-Sesquiphellandrene	Seed	Bhuiyan et al. (2009)
56	1-Undecanol	Leaf	Chung et al. (2012)
57	1-Dodecanol	Leaf, Stem	Chung et al. (2012)
58	Cyclododecanol	Leaf	Chung et al. (2012)
59	13-Tetradecenal	Leaf	Chung et al. (2012)
60	(Z)6-pentadecen-1-ol	Leaf	Chung et al. (2012)
61	2-Hexadecen-1-ol, 3, 7, 11, 15-tetramethyl-, [R-[R*, R*-(E)]]	Leaf	Chung et al. (2012)
62	2-(4-Methoxyphenyl) benzo[b]furan	Leaf	Chung et al. (2012)
63	9H-pyrrolo[3′,4′:3,4] pyrrolo[2,1-a]phthalazine-9,11(10H)-dione, 10-ethylphenyl4.	Stem	Chung et al. (2012)
64	1-Methyl-3-(3,4-dimethoxyphenyl)-6,7-dimethoxyisochromene	Stem	Chung et al. (2012)
65	Acrylic acid tetradecanyl ester	Stem	Chung et al. (2012)
66	Hexadecanoic acid, ethyl ester	Stem	Chung et al. (2012)
67	15-Methyltricyclo [6.5.2 (13,14),0(7,15)]pentadeca-1,3,5,7,9,11,13-heptene	Stem	Chung et al. (2012)
68	Linoleic acid ethyl ester	Stem	Chung et al. (2012)
69	Phytol	Stem	Chung et al. (2012)
70	Geraniol	Fruit	Momin et al. (2012)
71	cinnamic acid	Root	Tang et al. (2013)
72	4,4,5,7,8-pentamethyl-3,4-2 Isocoumarin-3-one,	Root	Tang et al. (2013)
73	1,3,4-tris(trimethylsilyloxy) octade-can-2-amine	Root	Tang et al. (2013)
74	L-valine	Root	Tang et al. (2013)

2.1.2. Seed

GC-MS analysis of petroleum ether extract of Korean variety of coriander revealed presence of various phytochemicals. Linalool, hotrienol, 3,7-Dimethylocta-1,7-dien-3,6-diol, 3,7-Octadiene-2,6-diol,2,6-dimethyl-, and 7-Oxabicyclo[4.1.0]heptanes,1-methyl-4-(2-methyloxiranyl)- were identified to be major phytochemicals (Nagella et al., 2012). Essential oils extracted from seeds of the coriander were analyzed for the presence of volatile compounds like linanool, γ-terpinene, α-pinene, camphor, decanal, geranyl acetate, limonene, geraniol, camphene, and D-limonene. It is observed that linanool was present around 55.5% by area in the gas chromatograph and it is regarded as a major constituent (Shahwar et al., 2012). From seed extracts of Korean coriander variety, various bioactive compounds were analyzed using ultra high-performance liquid chromatography. Flavonols like myricetin, quercetin, kaempferol and hydroxycinnamic acids like coumaric acids, chlorogenic acid and ferulic acid were isolated. Vanillin, rutin, pyrogallol, hesperidin, resveratrol, naringenin, formononetin, and veratric acid are some important phenolic acids isolated from Korean coriander variety. Along with these above-mentioned compounds various hydroxybenzoic acids like gallic acid, protocatechuic acid, β-resorcylic acid, vanillic acid, salicylic acid, gentisic acid, and syringic acid were isolated (Praveen et al., 2014). In cultivars of Bangladesh, the essential oils were extracted and analysed with GC-MS for the biochemical molecules. It is seen that linalool, geranyl acetate, γ-terpinene were present in high amounts (Bhuiyan et al., 2009). Some of the major phytochemicals reported from *Coriandrum sativum* L. is illustrated in Figure1.

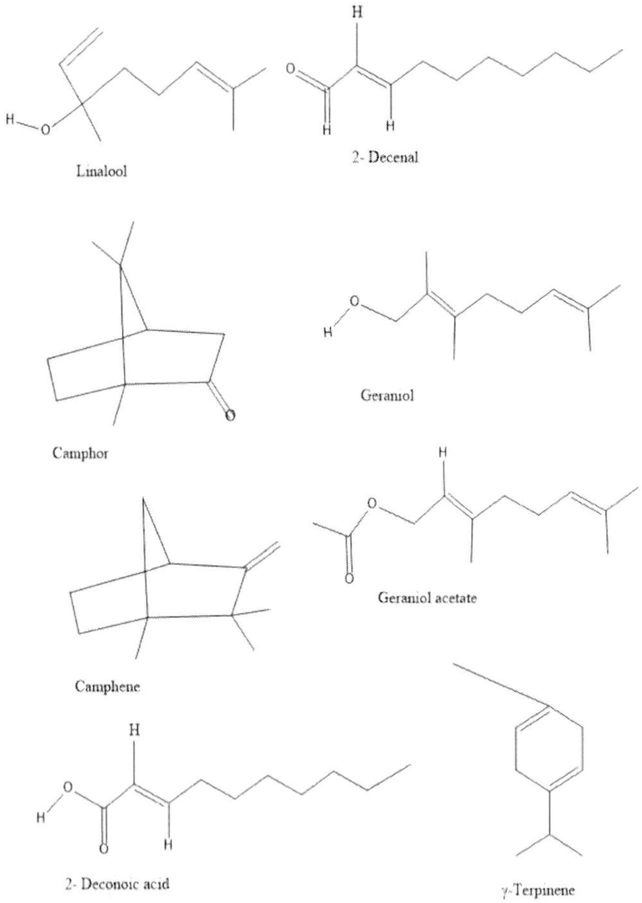

Linalool

2- Decenal

Camphor

Geraniol

Geraniol acetate

Camphene

2- Deconoic acid

γ-Terpinene

Figure 1. (Continued).

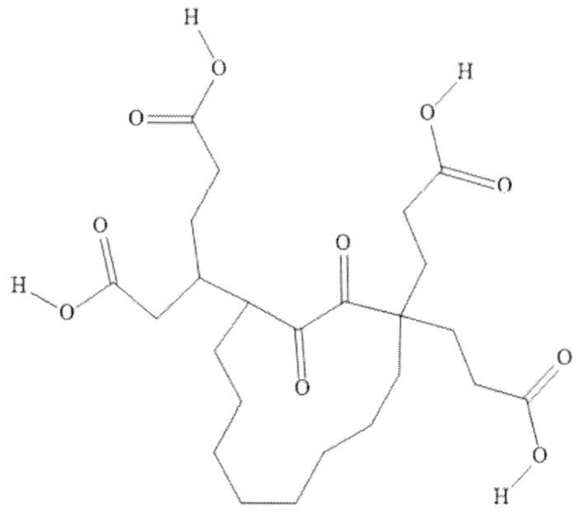

Plantaricin

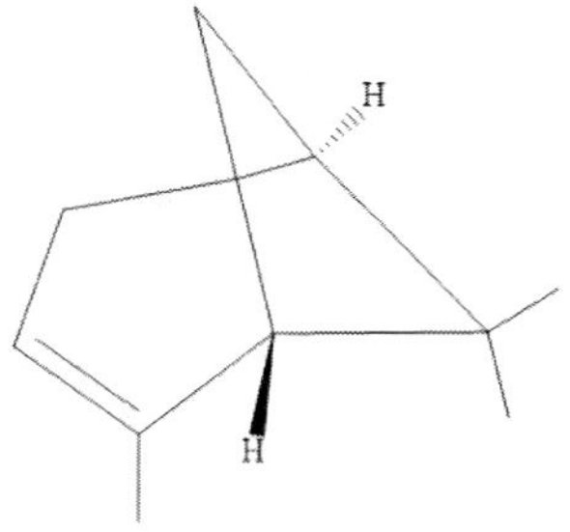

α Pinene

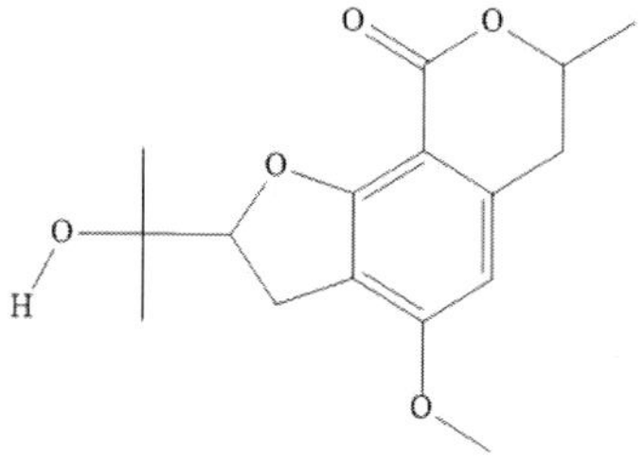

Coriandrone A

Source: https://pubchem.ncbi.nlm.nih.gov.

Figure 1. Major phytochemicals reported from *Coriandrum sativum* L.

2.1.3. Fruits

Fruits of coriander are rich in many biochemicals, they are rich in proteins, fats, crude fibres, starch etc., Essential oils extracted from fruits were analysed for biochemicals like linalool, which is the major component and monoterpene hydrocarbons like α- pinene, γ-terpine, geranyl acetate, camphor, and geraniol. Along with these components, there are different types of fatty acids present like petroselinic acid, linoleic acid, oleic acid, and palmitic acid. These components play a major role in contributing to the pharmacological properties (Momin et al., 2012). Different types of polyphenols like rutin, caffeic acid, feruloylquinic acid, and caffeoylquinic acid isomers were reported in fruits (Kaiser et al., 2013).

Table 3. Important Bioactive molecules isolated from *Coriandrum sativum* L.

S. No.	Bioactive molecules	Organ	Reference
1	Myricetin	Seed	Nagella et al. (2014); Kaiser et al. (2013)
2	Quercetin	Seed	Nagella et al. (2014)
3	Kaempferol	Seed	Nagella et al. (2014)
4	Chlorogenic acid	Seed	Nagella et al. (2014); Abascal and Yarnell (2012).
5	Ferulic acid	Seed	Nagella et al. (2014)
6	m-Coumaric acid	Seed	Nagella et al. (2014)
7	o-Coumaric acid	Seed	Nagella et al. (2014)
8	p-Coumaric acid	Seed	Nagella et al. (2014)
9	Gallic acid	Seed	Nagella et al. (2014)
10	Protocatechuic acid	Seed	Nagella et al. (2014)
11	Vanillic acid	Seed	Nagella et al. (2014)
12	Gentisic acid	Seed	Nagella et al. (2014)
13	β-Resorcylic acid	Seed	Nagella et al. (2014)
14	Salicylic acid	Seed	Nagella et al. (2014)
15	Syringic acid	Seed	Nagella et al. (2014)
16	Pyrogallol	Seed	Nagella et al. (2014)
17	Veratric acid	Seed	Nagella et al. (2014)
18	Vanillin	Seed	Nagella et al. (2014)
19	Hesperidin	Seed	Nagella et al. (2014)
20	Naringenin	Seed	Nagella et al. (2014)
21	Formononetin	Seed	Nagella et al. (2014)
22	Rutin	Seed	Nagella et al. (2014); Kaiser et al. (2013)
23	Resveratrol	Seed	Nagella et al. (2014)
24	Biochanin A	Seed	Nagella et al. (2014)
25	Coriandrone A	Leaf	Baba et al. (1991)
26	Coriandrone B	Leaf	Baba et al. (1991)
27	Coriandrone C	Leaf	Taniguchi et al. (1996)
28	Coriandrone D	Leaf	Taniguchi et al. (1996)
29	Coriandrone E	Leaf	Taniguchi et al. (1996)
30	β-carotene	Leaf, Stem	Guerra et al. (2005)
31	β-cryptoxanthin epoxide	Leaf, Stem	Guerra et al. (2005)
32	Lutein-5,6-epoxide	Leaf, Stem	Guerra et al. (2005)
33	Violaxanthin	Leaf, Stem	Guerra et al. (2005)
34	Neoxanthin	Leaf, Stem	Guerra et al. (2005)
35	Caffeic acid		Abascal and Yarnell (2012), Kaiser et al. (2013)
36	Gallic acid		Abascal and Yarnell (2012)
37	Feruloylquinic acid	Fruit	Kaiser et al. (2013)
38	Caffeoylquinic acid	Fruit	Kaiser et al. (2013)

2.1.4. Roots

Coriander roots were reported to have diversified phytochemicals like alkaloids, carbohydrates, flavonoids, saponins, sterols, and terpenoids (Kumar et al., 2014). Ethyl acetate extracts of coriander roots were analyzed for various phytochemicals using GC-MS. It is seen that phenolic compounds were reported to be highest and compounds like ascorbic acid, p-coumaric acid, cinnamic acid, Isocoumarin-3-one, 1,3,4-tris(trimethylsilyloxy)octadecan-2-amine, 4,4,5,7,8-pentamethyl-3,4-2 and L-valine were reported (Tang et al., 2013). The important bioactive molecules isolated from different parts of *Coriandrum sativum* L. has been tabulated in Table 3.

3. Medicinal Properties

Coriandrum sativum L., an annual herb, is globally being used in medicines for thousands of years. Besides its culinary use, it has potential medicinal benefits. The medicinal uses of coriander are documented in classical Greek and Latin literature (Manniche, 1989). Coriander has been used in treating digestive disorders, ulcers, halitosis, and rheumatism. It is a reservoir of essential oils which are produced in various parts of the plants, such as leaves, flower, fruit, and seed. They possess analgesic, anticancer, anticonvulsant, antidiabetic, anthelmintic, antihypertensive, antimicrobial, anti-mutagenic, antioxidant, anxiolytic, diuretic, hypnotic activity, and many more properties (Nadeem et al., 2013). Table 4 presents the phytochemicals in *Coriandrum sativum* L. that possess important pharmacological properties.

3.1. Antioxidant Activity

The polyphenolic extract of *C. sativum* exhibits antioxidant properties by inhibiting oxidative damage and by scavenging free radicals. It serves as an alternative to synthetic antioxidants. In the study by Sujatha and Srinivas, (1995), the peroxidized lipid-induced lysis is found to be inhibited by the aqueous extract of *C. sativum* seed (Sujatha and Srinivas, 1995). The catalase, glutathione peroxidase, glutathione reductase, and glutathione-S-transferase, superoxide dismutase activity increased and TBARS levels were decreased by treatment with polyphenolic fraction (Hashim et al., 2005). The antioxidant activity in coriander is ascribed to carotenoids (Guerra et al., 2005) and linalool, α-pinene, limonene, and camphene present in the essential oil (Wei and Shibamoto, 2007).

Table 4. Pharmacological activities of functional compounds in *Coriandrum sativum* L.

Pharmacological activity	Functional compounds	Reference
Antioxidant	carotenoids and linalool, α-pinene, limonene, and camphene	Guerra et al. (2005) Wei and Shibamoto (2007)
Antibacterial	Aliphatic (2E)-alkenals, alkanals, (2E)-Dodecenal, (2E) - undecenal, Plantaricin,	Zare-Shehneh et al. (2014)
Antifungal	Plantaricin	Zare-Shehneh et al. (2014)
Anthelmintic	linalool, p-mentha-1,4-dien-7-ol,α-pinene and neryl acetate	Zoubiri and Baaliouamer (2010)
Antihyperglycemic Hypolipidemic	Flavonoids and Polyphenols (Specific active compounds not detected)	Abderrahmane Aissaoui et al. (2011)
Cardioprotective	Vitamin E and ascorbic acid	Patel et al. (2012)
Hepatoprotective	Phenolic compounds	Pandey et al. (2011)
Diuretic	Flavonoids, caffeic acid derivatives, sesquiterpene lactones, triterpenes, coumarins, and carotenoids	Abderahim Aissaoui et al. (2008)
Antiulcer	Linalool, - terpinene and α- pinene in essential oils	Heidari et al. (2016)
Anxiolytic	Flavonoids and Linalool in essential oil	Mahendra and Bisht (2011) Emamghoreishi et al. (2005)
Anticancer	Linalool (specific compound for anticancer activity is still under research)	Elmas et al. (2019)
Antimutagenicity	Chlorophyll	Cortés-Eslava et al. (2004)
Immunomodulatory	Linalool in essential oils, petroselinic acid, glycitein, pyrogallol, and caffeic acid	Ahmed et al. (2020)
Metal detoxification	Kaempferol and acacetin flavonoids	Winarti et al. (2018)

3.2. Antimicrobial Activity

The essential oil in coriander has shown to possess antimicrobial activity against a wide range of pathogens. The coriander extract and the essential oil exhibits both antibacterial and antifungal activity.

3.3. Antibacterial Activity

The seed and leaf extracts and the oil of *C. sativum* have shown antibacterial activity against gram-positive and gram-negative bacteria such as *Bacillus megaterium*, *Escherichia coli*, *Enterococcus faecalis*, *Klebsiella pneumoniae*, *Listeria monocytogenes*, *Pseudomonas aeruginosa*, *Salmonella typhimurium*, *Staphylococcus aureus*, *Yersinia enterocolitica* and many other bacterial strains (Delaquis et al., 2002); (Keskin and Toroglu, 2011). The aliphatic (2E)-alkenals, alkanals, (2E)-Dodecenal, (2E) - undecenal characterized in the coriander have shown bactericidal activity against the microbes. Plantaricin, an antimicrobial peptide isolated from coriander leaf extract exerted antimicrobial activity against *K. pneumoniae*, *P. aeruginosa*, and *S. aureus* bacteria with minimum inhibitory concentration (MIC) values of 71.55, 86.4, and 35.2 μg/mL respectively (Zare-Shehneh et al., 2014). The plant has induced disease resistance in catla catla fish infected with *Aeromonas hydrophila* by enhancing immunostimulant potential (Innocent, 2011).

3.4. Antifungal Activity

The essential oil of *C. sativum* (CEO) has served as a potent fungicide against fungi that deteriorate the high moisture foods. The use of 0.15% of CEO in cakes has controlled the growth of molds like *Aspergillus niger*, *Monilia sitophila*, *Penicillium expansum*, *Penicillium stoloniferum*, *Rhizopus stolonifer*, etc (Darughe et al., 2012). The minimal fungicidal concentration (MFC) of CEO and active fraction ranges from 31.2–62.5 μg/mL and 125 μg/mL to 1000 μg/mL, respectively for Candida spp. (Freires et al., 2014). The seed-borne pathogens in paddy like *Alternaria alternata*, *Bipolaris oryzae*, *Curvularia lunata*, *Drechslera halodes*, *Fusarium oxysporum*, *Tricoconis padwickii* are inhibited by the application of coriander oil (Lalitha et al., 2011). Plantaricin in the *C. sativum* has shown antifungal property against *A. niger* and *Penicillium lilacinum* with MICs 62.1 and 67.8 μg/mL respectively (Yogesh et al., 2014).

3.5. Anthelmintic Activity

Parasitic worms pose a threat to the food industry by infecting food crops and livestock. Anthelmintic drugs are those which flush out the parasites from the body by killing them (Chandan et al., 2011). As per WHO synthetic drugs like Albendazole, ivermectin are used (Onakpoya 2018). *C. sativum* with many therapeutic properties is known to be effective against parasitic worms. The crude aqueous and hydro-alcoholic extract of *C. sativum* fruits were analysed for its anthelmintic activity against the egg and adult nematode *Haemonchus contortus*. Both the extracts inhibited the hatching of eggs at a concentration of 0.5 mg/ml (Eguale et al., 2007). The essential oil has shown toxicity against *Sitophilus granarius, a* pathogen in chickpea grains (Zare-Shehneh and Baaliouamer, 2010).

3.6. Antidiabetic/Antihyperglycemic

C. sativum has been used as an herbal medicine for treatment of diabetes mellitus, a metabolic disorder marked by hyperglycemia. The seeds and fruits of coriander have shown antidiabetic activity by stimulating insulin secretion and enhancing the glucose uptake and metabolism by muscle. The coriander fruit-extract suppressed hyperglycemia in hyperglycemic Meriones shawi rats by decreasing the insulin resistance (Abderrahmane Aissaoui et al., 2011). In another study, the aqueous consumption increased insulin secretion, 2-deoxyglucose transport and glucose incorporation into glycogen by 1.3-5.7 folds, 1.6 folds and 1.7 folds respectively (Gray and Flatt, 1999). In a study conducted by Abascal and Yarnell (2012), the fasting blood-sugar levels decreased in the volunteers who consumed 2.5 g of ground coriander fruit for 60 days twice daily. Thus, coriander serves as a potential antihyperglycemic oral dietary supplement.

3.7. Hypolipidemic

Hyperlipidemia is a condition in which there is a high level of fat generation and accumulation in subendothelial spaces of bone and vasculature. The hypolipidemic effect of coriander extract has been shown in rats with induced hyperlipidemia. The coriander extract at the dosage of 1g/kg reduced the uptake of lipids and enhanced the breakdown of lipids (Lal et al., 2004). In another study, the total triglycerides and cholesterol levels decreased significantly and β-hydroxy, β-methyl glutaryl CoA reductase fat-fed animals treated with coriander seeds (Chithra and Leelamma, 1997).

3.8. Cardioprotective Effects

High blood pressure is one of the most common cardiovascular disorders and coriander protects the heart by lowering the blood pressure and reducing bad cholesterol. The cardiac damage has been prevented by the administration of hydro-methanolic extract of coriander fruits in Wistar rats with isoproterenol-induced cardiotoxicity. The methanolic extract inhibited myofibrillar damage by reducing myocardial infarction in rats (Patel et al., 2012). The coriander fruits have shown to decrease HDL cholesterol, cholesterol-associated lipids, and dyslipidemia. The extract has improved the cardioprotective indices and blood-fat values, thus showing a potential cardioprotective effect (Abascal and Yarnell, 2012).

3.9. Hepatoprotective Effects

Hepatotoxicity is one of the leading liver diseases which accounts for 15% of the world's disease. Hepatoprotective drugs are used to treat hepatotoxicity. Natural-derived hepatotoxic drugs are prioritized (Kumar et al., 2011). Coriander benefits the normal functioning of the liver. The *C. sativum* extract protects the liver from oxidative stress and serves as a powerful antioxidant by means of linalool in the essential oil. The study conducted by Sreelatha et al., (2009) has shown that the rats with oxidative stress induced by carbon tetrachloride when orally administered with the coriander extract reduced SGOT, SGPT, TBARS, SOD, catalase activity. The hepatoprotective activity is comparable with silymarin, a standard drug.

3.10. Diuretic Effects

A diuretic is a substance that promotes diuresis by enhancing the excretion of water and from the body during conditions like high blood pressure, heart failure, kidney related issues. Medicinal plants are a major source of herbal diuretics which stands out to be most popular in recent times. The aqueous extract of coriander administered at the dosage of 40 and 100 mg/kg in anesthetized Wistar rats. Total urine excreted out was collected and volume was calculated. It has increased the excretion of chloride, potassium, and sodium in turn promoting diuresis. The diuretic activity of coriander aqueous extract is comparable to that of furosemide in the excretion of electrolytes, and glomerular filtration (Abderahim Aissaoui et al., 2008; Jabeen et al., 2009).

3.11. Anxiolytic Effects

Anxiety is a common disorder associated with worse quality of lifestyle and poorer neuropsychological performance. Coriander has been used as an anti-anxiety agent in traditional medicine. The anxiolytic activity of hydroalcoholic extract of coriander has been studied in mice using diazepam as standard by Mahendra and Bisht (2011) and it was found that the anti-anxiety effect of coriander extract was similar to diazepam at the concentration of 100 and 200 mg/kg. In a study by Emamghoreishi et al., (2005), the aqueous extract of coriander showed potential anxiolytic activity at 50, 100, and 500 mg/kg. In an adult man (75 kg), 7.5 g coriander fruit dry extract is suggested as an effective dose (Önder, 2018).

3.12. Antiulcer Effects

Peptic ulcers are lesions in the gastrointestinal tract due to disturbance in the mucosal resistance affecting nearly 10% of the world's population. There are many synthetic drugs available in the market to treat peptic ulcers. The medicinal plants rich in therapeutic secondary metabolites are used to treat peptic ulcers (Vimala and Gricilda Shoba, 2014). Coriander is used as an ayurvedic remedy for open sores that develop in the inner lining of the mouth and stomach. The seeds of the herb are effective against *Helicobacter pylori* that causes peptic ulcers and thus prevents the formation of gastric ulcers. The antioxidant property of *C. sativum* is correlated with antigastric activity. The plant extract scavenges reactive oxygen species on the surface of gastric mucosa. The coriander extract also forms a protective hydrophilic layer against gastric injury (Önder, A. 2018). In the study by Abascal and Yarnell (2012), the fruit extracts of coriander at the dosage of 250 mg/kg showed antiulcerogenic effects against ethanol, indomethacin, and sodium hydroxide.

3.13. Anti-Aging Effects

The long chain fatty acid forms an essential part of epidermis of skin inturn serving as an anti-aging agent. Most of the anti-aging products available in the market contain long chain fatty acids. Coriander essential oil is a rich source of long chain fatty acids and can be used as a topical treatment for skin healing and they are excellent alternatives for sunscreen. Umbelliferin an extract obtained from coriander fruit contains petroselinic acid triglycerides aids in supporting skin protection (Majeed and Prakash, 2005). The skin gets photodamaged on exposure to UV rays increasing collagen production and matrix metalloproteinase-1 (MMP-1) activity in turn causing skin wrinkles. In a study by Hwang et al., (2014) it has been shown that the coriander extract increases procollagen type I production, decreases the MMP-1 levels in the normal human dermal fibroblasts (NHDF), exposed to UV rays. The NHDF cells also showed dense dermal fibres and thinner epidermal layers than those which were untreated with coriander extract.

3.14. Anticancer Effects

Herbal extracts have been serving as potential anticancer agents with less side effects. Coriander fruit extract is one such herbal medicine that prevents malignancy and the development of cancer. The study on the anticancer effects of coriander has been conducted in colon cancer-induced rats induced by dimethylhydrazine. The coriander fruit extract has reduced cholesterol and increased the bile acids and fecal-neutral sterols, thereby normalizing the lipid metabolism in colon cancer (Chithra and Leelamma, 2000). The effect of coriander extract has been studied in prostate cancer cells and it is found to inhibit malignancy and suppress the expression of genes associated with cancer (Ware et al., 2019). The effect of CSE extract was studied in PC-3 and LNCaP prostate cancer cell lines. The extract inhibited the colony formation, cell invasion and migration in PC-3 prostate cell lines but not in the LNCaP cell lines (Elmas et al., 2019). The root, stem and leaf extract are also found to be effective against breast cancer (Ware et al., 2019). The CSE has been assessed for its antiproliferative effects in breast cancer cell line, MCF-7 using MTT assay. It was found that the CSE showed anticancer activity in MCF-7 cells by arresting the cell cycle at G2/M phase, causing apoptotic cell death and by affecting the antioxidant enzymes (Tang et al., 2013).

3.15. Antimutagenicity Activity

Antimutagens are compounds that decrease or nullify the mutagenic effects of harmful agents. A number of plants with antimutagenic principles such as *Smilax china*, *Prunella vulgaris* and *Pteris multifida* are studied for their effect against mutagens and many herbal drugs have been developed as antimutagens to reverse multistage carcinogenesis (Akram et al., 2020). *C. sativum* are known to prevent the transformation of mutagenic compounds to mutagen thus serving as an effective antimutagen. Ample research has been conducted on the antimutagenic effect of coriander. Ame's test is employed in the assessment of the antimutagenic effects of the plant against mutagens like 4-nitro-o-phenylenediamine, m-phenylenediamine and 2-aminofluorene in the indicator organism *S. typhimurium* TA98. It has been observed that the mutagenic effect of three tested amines has been reduced by coriander crude extract. Also the extract is known to induce the expression of DNA damage responsive genes in response to RNR3 and RAD51 (Cortés-Eslava et al., 2004; Bakkali et al., 2005).

3.16. Immunomodulatory Activity

Immunomodulatory drugs are a class of drugs that modify the immune response by increasing or decreasing the antibodies. They are classified as immunosuppressants and immunostimulants. Immunostimulants enhance the immune response against infectious diseases and immunodeficiency by enhancing antibody production while immunosuppressants suppress the immune response against newly transplanted organs and are used in the treatment of autoimmune diseases (Bascones-Martinez et al., 2014). Most of these immunomodulators are known to cause adverse side effects. Thus, there is rapid demand for herbal medicine that modulates the immune system (Jantan et al., 2015). Coriander is enriched with antioxidants which boosts the immune system. It has been reported that the crude extract of coriander proliferates human peripheral blood mononuclear cells (PBMC) and the secretion of interferons like IFN-γ by Cherng et al., (2008). In another study, the immunomodulatory activity of coriander seed powder and extract in *Oreochromis niloticus* L. has been assessed and is found to ameliorate the immunosuppressive effects induced by lead (Ahmed et al., 2020).

3.17. Metal Detoxification

Heavy metals can have detrimental effects when they accumulate beyond a limited range in the body. The heavy metals enter the body through food or water. These heavy metals modify the chemical structures modifying and disturbing its normal biological functioning (Aprile and De Bellis, 2020). The heavy metal detox is very important to overcome the ill effects caused by metal. Plant extracts can serve as effective metal detoxifiers and coriander serves as a potent tool for detoxification by aiding the excretion of metals naturally as it has chelating properties. The amphoteric electrolytes, citric acid, and phytic acid bind to metals and help in metal sequestration (Mehrandish et al., 2019). The methanolic extract of the *Coriandrum sativum* (50mg/kg) was administered into male Wistar rats poisoned with lead acetate (50mg/kg) to check the metal detoxification effects. It was found that the lead concentration was decreased in the treated groups and showed lesser histological damage (Téllez-López et al., 2017).

4. Cultivation and Its Processing

Coriandrum sativum is an annual plant that grows in warm weather. A frost-free tropical environment favors the growth of coriander. It is grown during spring and autumn for its seeds and leaves. The seeds are sown in the month of June to July and October to November. Loamy soil or well-drained silt that retains moisture is suitable for the cultivation of coriander. The soil that is rainfed has to be in the form of clay (Reddy, 2015). The pH of 6 – 8 and a temperature range of 20–25°C are optimum for the growth of coriander. The seeds have to be hardened by treatment with potassium dihydrogen phosphate for 16 hours followed by *Azospirillum* and *Trichoderma viride* to control wilt disease prior to sowing. Soil has to be fertilized with 10 kg nitrogen, 40 kg phosphate, and 20 kg potassium (Yogesh et al., 2017). The seeds are then sown at the depth of ¼ to ½ inch and placed at a distance of 12-15 inches (Masabni, 2016). They germinate within 10-14 days after sowing. It requires 5-6 irrigations. They can be harvested when they are green or left until the fruits are ripe and turn brown. The plants are harvested at 30-40 days after sowing for leaves and are harvested at about 90 to 110 days when the fruits are fully ripe for the seeds.

The coriander is generally consumed in unprocessed form. However, it can be processed for long-term preservation or for international trade. The coriander plants can be subjected to vacuum cooling immediately after harvest to prolong shelf life (Apichart et al., 2012). The microbial contamination can be reduced by irradiating it with a gamma radiation dose of 0.5 kGy (Cruz-Zaragoza et al., 2011). The whole plant is sun dried after harvesting until the moisture content drops to 18%. The seeds are then separated by light threshing and winnowing and further dried in shade to reduce the moisture content to

9%. The other means of drying are freeze-drying and microwave drying etc. The seeds are crushed and subjected to steam distillation for extraction of essential oils (Yogesh et al., 2017). The essential oils can also be extracted by using supercritical CO_2 extraction using carbon dioxide and propane as solvents with a CO_2 pressure of 200 and 300 bar and temperature of 35°C. In the study conducted by Illes et al., (2000), the highest extraction of essential oils has been achieved by using propane solvents at 25°C and CO_2 pressure at 50, 80, and 100 bar (Illés et al., 2000). The essential oil can be encapsulated in calcium alginate (sodium alginate and calcium chloride) to prevent the loss of aromatic flavor and nutrition. The leaves can be processed into purees, paste, and sauce. The leaves can be powdered after blanching at 90°C for 2 minutes to prevent the peroxidase activity and preserved for further use. The leaves and fruits can be preserved in the form of the paste after steam blanching at 100°C or water blanched at 90 and 100°C for 1-10 minutes. When checked for the antioxidants and phenolic contents by HPLC, the levels remained unchanged with steam blanching for one minute. However, increased levels were observed with water blanching and extended steam blanching and it was even higher with short time water blanching (Kaiser et al., 2013).

Coriandrum sativum is produced either on a small scale or large scale in majority of the countries worldwide. Most of the coriander produced is used up in the local markets. With proper cultivation and farming practices, an average yield of coriander is expected to be 20 quintals per acre. The current market rate per quintal as specified by NCDEX is 6800 INR that will sum up to 1,36,000 INR per acre (Banerjee, 2020).

Conclusion

Coriander is a phenomenal herb that has a dual function as a spice as well as an herbal medicine. It is an annual herb that can be easily cultivated and can be grown throughout the year under controlled conditions with least investment on its cultivation. Coriander is rich in nutrition with fat, proteins, minerals, vitamins etc. The essential oil present in the plant has a broad range of applications in human health, agriculture, and the environment. It can be used in cosmetic, food and pharmaceutical industries. The phytochemicals in the plant have potential health benefits and possess analgesic, anticancer, anticonvulsant, antidiabetic, anthelmintic, antihypertensive, antimicrobial, anti-mutagenic, antioxidant, anxiolytic, diuretic, hypnotic activity, and many more properties. Thus, it is highly recommended to include coriander in the diet due to its multifunctional activities and protective action against various diseases. Moreover, long term preservation of the herb can be achieved by adopting new processing methods or improvising the traditional processing methods. Extensive research needs to be done to discover the undiscovered active compounds and the therapeutic potential of this plant.

References

Abascal, K., and E. Yarnell, 2012. "Cilantro—Culinary Herb or Miracle Medicinal Plant." *Alternative and Complementary Therapies*.18: 259–264.

Ahmed, S. A. A., R. M. Reda, and M. ElHady, 2020. "Immunomodulation by *Coriandrum sativum* seeds (Coriander) and its ameliorative effect on lead-induced immunotoxicity in Nile tilapia (*Oreochromis niloticus* L.)." *Aquaculture Research.* 51:1077-1088.

Aissaoui, A., J. El-Hilaly, Z. H. Israili, and B. Lyoussi, 2008. "Acute diuretic effect of continuous intravenous infusion of an aqueous extract of *Coriandrum sativum* L. in anesthetized rats." *Journal of Ethnopharmacology*. 115:89–95.

Aissaoui, A., S. Zizi, Z. H. Israili, and B. Lyoussi, 2011. "Hypoglycemic and hypolipidemic effects of *Coriandrum sativum* L. in Meriones shawi rats." *Journal of Ethnopharmacology*.137:652–661.

Akram, M., M. Riaz, A. W. C. Wadood, A. Hazrat, M. Mukhtiar, S. Ahmad Zakki, R. Zainab, 2020. "Medicinal plants with anti-mutagenic potential." *Biotechnology & Biotechnological Equipment*. 34:309–318.

Apichart, S., B. Danai, and B. Pichaya, 2012. "Effect of Vacuum Cooling on Physico-chemical Properties of Organic Coriander." *Asian Journal of Food and Agro-Industry*. 5:96-103.

Baba, K., Y.-Q. Xiao, M. Taniguchi, H. Ohishi, and M.Kozawa, 1991. "Isocoumarins from *Coriandrum sativum*." *Phytochemistry*. 30:4143–4146.

Bakkali, F., S. Averbeck, D. Averbeck, A. Zhiri, and M. Idaomar, 2005. "Cytotoxicity and gene induction by some essential oils in the yeast *Saccharomyces cerevisiae*." *Mutation Research*. 585:1–13.

Banerjee, A. (n.d.). *Complete Guide to Coriander Farming: Varieties, Climate Requirement, Harvesting, Yield and Economics*. [online]. 2020 [cited 2021 May 30] Available from https://krishijagran.com/agripedia/complete-guide-to-coriander-farming-varieties-climate-requirement-harvesting-yield-and-economics/.

Bascones-Martinez, A., R. Mattila,R. Gomez-Font,and J. H. Meurman, 2014. "Immunomodulatory drugs: oral and systemic adverse effects." *Medicina Oral Patologia Oral Y Cirugia Bucal.* 19:e24–31.

Bhuiyan, M. N. I., J. Begum,and M. Sultana, 2009. "Chemical composition of leaf and seed essential oil of *Coriandrum sativum* L. from Bangladesh." *Bangladesh Journal of Pharmacology*. 4:150-153.

Burdock, G. A., andI. G. Carabin, 2009. "Safety assessment of coriander (*Coriandrum sativum* L.) essential oil as a food ingredient." *Food and Chemical Toxicology:*

AnInternational Journal Published for the British Industrial Biological Research Association. 47:22–34.

Chandan, H. S., A. R. Tapas, and D. M. Sakarkar, 2011. "Anthelmintic activity of extracts of *Coriandrum sativum* linn. in indian earthworm." *Phytomedicine: International Journal of Phytotherapy and Phytopharmacology*. 3:36-40.

Cherng, J. M., W. Chiang, and L. C. Chiang, 2008. "Immunomodulatory activities of common vegetables and spices of Umbelliferae and its related coumarins and flavonoids." *Food Chemistry*. 106:944–950.

Chithra, V., and S. Leelamma, 1997. "Hypolipidemic effect of coriander seeds (*Coriandrum sativum*): mechanism of action." *Plant Foods for Human Nutrition*. 51:167–172.

Chithra, V., and S. Leelamma, 2000. "*Coriandrum sativum* — effect on lipid metabolism in 1,2-dimethylhydrazine induced colon cancer." *Journal of Ethnopharmacology*. 71:457–463.

Chung, I. M., A. Ahmad, S. J. Kim, P. M. Naik, and P. Nagella, 2012. "Composition of the essential oil constituents from leaves and stems of Korean *Coriandrum sativum* and their immunotoxicity activity on the *Aedes aegypti* L." *Immunopharmacology and immunotoxicology*. 34:152-156.

Cortés-Eslava, J., S.Gómez-Arroyo, R. Villalobos-Pietrini, and J. J. Espinosa-Aguirre, 2004. "Antimutagenicity of coriander (*Coriandrum sativum*) juice on the mutagenesis produced by plant metabolites of aromatic amines." *Toxicology Letters*.153:283–292.

Cruz-Zaragoza, E., B. Ruiz-Gurrola, C. Wacher, T. Flores Espinosa, and M.Barboza-Flores, 2011. "Gamma radiation effects in coriander (*Coriandrum sativum* L) for consumption in Mexico."*Revista Mexicana de Física*. 57:80–86.

Darughe, F., M.Barzegar, and M. A. Sahari, 2012. "Antioxidant and antifungal activity of Coriander (*Coriandrum sativum* L.) essential oil in cake." *Food and Chemical Toxicology: An International Journal Published for the British Industrial BiologicalResearch Association*. 19:1253–1260.

Delaquis, P. J., K. Stanich, B. Girard, and G. Mazza, 2002. "Antimicrobial activity of individual and mixed fractions of dill, cilantro, coriander and eucalyptus essential oils." *International Journal of Food Microbiology*.74:101-109.

Diederichsen, A., and K. Hammer, 2003. "The infraspecific taxa of coriander (*Coriandrum sativum* L.)." *Genetic Resources and Crop Evolution*. 50:33–63.

Diederichsen, A., 1996. Taxonomy and names of the species. In International Plant Genetic Resources Institute. (Eds.), *Coriander: Coriandrum Sativum L. Bioversity International*. (3, pp. 8-10).Italy, Rome: The Alliance of Bioversity International and CIAT.

Eguale, T., G. Tilahun, A. Debella, A. Feleke, and E. Makonnen, 2007. "In vitro and in vivo anthelmintic activity of crude extracts of *Coriandrum sativum* against *Haemonchus contortus*." *Journal of Ethnopharmacology*. 110:428–433.

Elmas, L., M. Secme, R. Mammadov, U. Fahri Oglu, and Y. Dodurga, 2019. "The determination of the potential anticancer effects of *Coriandrum sativum* in PC-3 and LNCaP prostate cancer cell lines." *Journal of Cellular Biochemistry*. 120: 3506–3513.

Emamghoreishi, M., M.Khasaki, and M. F. Aazam, 2005. "*Coriandrum sativum*: evaluation of its anxiolytic effect in the elevated plus-maze." *Journal of Ethnopharmacology*.96:365–370.

Freires, I. de A., R. M. Murata, V. F. Furletti, A. Sartoratto, S. M. de Alencar, G. M. Figueira, J. A. de Oliveira Rodrigues,M. C. T. Duarte,and P. L. Rosalen, 2014. "*Coriandrum sativum* L. (Coriander) essential oil: antifungal activity and mode of action on Candida spp., and molecular targets affected in human whole-genome expression" *PloS One*. 9:e99086.

Gray, A. M., and P. R. Flatt, 1999. "Insulin-releasing and insulin-like activity of the traditional anti-diabetic plant *Coriandrum sativum* (coriander)." *The British Journal of Nutrition*. 81:203–209.

Guerra, N. B., E. de Almeida Melo, and J. M. Filho, 2005. "Antioxidant compounds from coriander (*Coriandrum sativum* L.) etheric extract." Journal of Food Composition and Analysis: An Official Publication of the United Nations University, *International Network of Food Data Systems*. 18:193–199.

Hwang, E., D. G. Lee, S. H. Park,M. S. Oh,and S. Y. Kim, 2014. "Coriander leaf extract exerts antioxidant activity and protects against UVB-induced photoaging of skin by regulation of procollagen type I and MMP-1 expression." *Journal of Medicinal Food*. 17: 985–995.

Hashim, M. S., S. Lincy, V. Remya, M. Teena, and L. Anila, 2005. "Effect of polyphenolic compounds from *Coriandrum sativum* on H_2O_2-induced oxidative stress in human lymphocytes." *Food Chemistry*. 92:653–660.

Heidari, B., S. E. Sajjadi, and M. Minaiyan, 2016. "Effect of *Coriandrum sativum* hydroalcoholic extract and its essential oil on acetic acid- induced acute colitis in rats." *Avicenna Journal of Phytomedicine*. 6:205–214.

Illés, V., H. G. Daood, S. Perneczki, L. Szokonya, and M. Then, 2000. "Extraction of coriander seed oil by CO_2 and propane at super- and subcritical conditions." *The Journal of Supercritical Fluids*.17:177–186.

Innocent, B. X., A.F.M. Syed, and Dhanalakshmi, 2011. "Studies on the immunostimulant activity of *Coriandrum sativum* and resistance to *Aeromonas hydrophila* in Catla catla." *Journal of Applied Pharmaceutical Science*. 1:132-135.

Iqbal, M. J., M. S. Butt, and H. A. R. Suleria, 2017. Coriander (*Coriandrum sativum* L.): bioactive molecules and health effects. J.M. Mérillon and K.G. Ramawat (Eds.), *Bioactive Molecules in Food*, 1–37. Switzerland: Springer.

Jabeen, Q., S.Bashir, B. Lyoussi, and A. H. Gilani, 2009. "Coriander fruit exhibits gut modulatory, blood pressure lowering and diuretic activities." *Journal of Ethnopharmacology* 122:123–130.

Jantan, I., W. Ahmad, and S. N. A. Bukhari, 2015. "Plant-derived immunomodulators: an insight on their preclinical evaluation and clinical trials." *Frontiers in Plant Science*. 6:655.

Kaiser, A., D. R. Kammerer, and R. Carle, 2013. "Impact of blanching on polyphenol stability and antioxidant capacity of innovative coriander (*Coriandrum sativum* L.) pastes." *Food Chemistry*. 140:332–339.

Keskin, D., and S. Toroglu, 2011. "Studies on antimicrobial activities of solvent extracts of different spices." *Journal of Environmental Biology*.32:251–256.

Kumar, C. H., A. Ramesh, and J. N. S. Kumar, 2011. "A review on hepatoprotective activity of medicinal plants." *International Journal of Pharmaceutical Sciences and Research.* 4:501-515.

Kumar, R. S., P. Balasubramanian, P. Govindaraj, and T. Krishnaveni, 2014. "Preliminary studies on phytochemicals and antimicrobial activity of solvent extracts of *Coriandrum sativum* L. roots (Coriander)." *Journal of Pharmacognosy and Phytochemistry*. 2:74-78.

Lal, A. A. S., Kumar, T., P. B. Murthy, and K. S. Pillai, 2004. "Hypolipidemic effect of *Coriandrum sativum* L. in triton-induced hyperlipidemic rats." *Indian Journal of Experimental Biology*.42:909–912.

Lalitha, V., B. Kiran, and K. A. Raveesha, 2011. Antifungal and antibacterial potentiality of six essential oils extracted from plant source." *International Journal of Engineering Science Technologies*. 3:3029–3038.

Laribi, B., K. Kouki, M. M'Hamdi, and Bettaieb, T., Bettaieb, 2015. "Coriander (*Coriandrum sativum* L.) and its bioactive constituents." *Fitoterapia*.103:9-26.

Mahendra, P., and S. Bisht, 2011. "Anti-anxiety activity of *Coriandrum sativum* assessed using different experimental anxiety models." *Indian Journal of Pharmacology.* 43:574–577.

Majeed, M., and L. Prakash, 2005. "*Novel natural approaches to anti-aging skin care.*" Cosmetics and Toiletries Manufacture Worldwide. 1:11–15.

Mandal, S., and M. Mandal, 2015. "Coriander (*Coriandrum sativum* L.) essential oil: Chemistry and biological activity." *Asian Pacific Journal of Tropical Biomedicine.* 5:421–428.

Manniche, L. 1989. *The Herbal. An ancient Egyptian herbal.* (pp. 61-158). USA, Texas: University of Texas Press.

Masabni, J. 2016. *Growing cilantro*.[online]. 2016 cited (2021, April 15) Available from: URL: https://agrilifeextension.tamu.edu/library/gardening/cilantro/.

Mehrandish, R., A. Rahimian, and, A.Shahriary, 2019. Heavy metals detoxification: "A review of herbal compounds for chelation therapy in heavy metals toxicity." *Journal of Herbmed Pharmacology*.8:69–77.

Momin, A. H., S. S. Acharya, and A. V. Gajjar, 2012. "*Coriandrum sativum*-review of advances in phytopharmacology." *International Journal of Pharmacy and Pharmaceutical Sciences*. 3:1233.

Nadeem, M., F. Muhammad Anjum, M. Issa Khan, S. Tehseen, A. El-Ghorab, and J. Iqbal Sultan, 2013. Nutritional and medicinal aspects of coriander (*Coriandrum sativum* L.): A review. *British Food Journal*. 115:743–755.

Nagella, P., M. Y. Kim, A. Ahmad, M. Thiruvengadam, and I. M. Chung, 2012. "Chemical constituents, larvicidal effects and antioxidant activity of petroleum ether extract from seeds of *Coriandrum sativum* L." *Journal of Medicinal Plants Research*. 6:2948-2954.

Nagella P., M.Thiruvengadam, A. Ahmad, and I.M. Chung, 2014. "UHPLC analysis of polyphenol composition and antioxidant activity from different solvent extracts of *Coriandrum sativum* seeds cultivated in Korea." *Asian Journal of Chemistry*. 26:6351-6356.

Önder, A. 2018. Coriander and Its Phytoconstituents for the Beneficial Effects. In H. A. El-Shemy (Ed.), *Potential of Essential Oils*. IntechOpen.

Onakpoya, I. J. 2018. Antihelminthic Drugs. In S. D. Ray (Ed.), *Side Effects of Drugs Annual* (40, pp. 377–382). Elsevier.

Pandey, A. P. Bigoniya, V.Raj, and K. K. Patel, 2011. "Pharmacological screening of *Coriandrum sativum* Linn. for hepatoprotective activity." *Journal of Pharmacy & Bioallied Sciences*. 3:435–441.

Patel, D. K., S. N. Desai, H. P. Gandhi, R. V. Devkar, and A. V. Ramachandran, 2012. "Cardio protective effect of *Coriandrum sativum* L. on isoproterenol induced myocardial necrosis in rats." *Food and Chemical Toxicology: An International Journal Published for the British Industrial Biological Research Association*. 50:3120–3125.

Potter, T. L., and I. S. Fagerson, 1990. "Composition of coriander leaf volatiles." *Journal of Agricultural and Food Chemistry*. 38:2054–2056.

Ramezani, S., F. Rasouli, and B. Solaimani, 2009. "Changes in essential oil content of coriander (*Coriandrum sativum* L.) aerial parts during four phonological stages in Iran." *Journal of Essential Oil Bearing Plants*. 12:683–689.

Shahwar, M. K., A. H. El-Ghorab, F. M. Anjum, M. S. Butt, S. Hussain, and M. Nadeem, 2012. "Characterization of Coriander (*Coriandrum sativum* L.) Seeds and Leaves: Volatile and Non Volatile Extracts." *International Journal of Food Properties*. 15:736–747.

Silva, F., C. Domeño, and F. C. Domingues, 2020. *Coriandrum sativum* L.: Characterization, Biological Activities, and Applications V. R. Preedy and WatsonIn R. R. (Eds.), *Nuts and Seeds in Health and Prevention* (pp. 497–519). USA: Academic Press.

Sreelatha, S., P. R. Padma, and M. Umadevi, 2009. "Protective effects of *Coriandrum sativum* extracts on carbon tetrachloride-induced hepatotoxicity in rats." *Food and Chemical Toxicology: An International Journal Published for the British Industrial Biological Research Association*. 47:702–708.

Sujatha, R., and L. Srinivas, 1995. "Modulation of lipid peroxidation by dietary components." *Toxicology in Vitro: An International Journal Published in Association with BIBRA*. 9:231–236.

Tang, E. L. H., J. Rajarajeswaran, S. Y. Fung, and M. S. Kanthimathi, 2013. "Antioxidant activity of *Coriandrum sativum* and protection against DNA damage and cancer cell migration." *BMC Complementary and Alternative Medicine*. 13:347.

Taniguchi, M., M. Yanai, Y. Q. Xiao, T.-I. Kido, and K. Baba, 1996. "Three isocoumarins from *Coriandrum sativum*." *Phytochemistry*. 42:843–846.

Téllez-López, M. Á., G. Mora-Tovar, I. M. Ceniceros-Méndez, C. García-Lujan, C. O. Puente-Valenzuela, M. D. C. Vega-Menchaca, J. Morán-Martínez, 2017. "Evaluation of the chelating effect of methanolic extract of *Coriandrum sativum* and its fractions on Wistar rats poisoned with lead acetate." *African Journal of Traditional, Complementary, and Alternative Medicines*. 14:92–102.

Ware, M., RDN, and D, L. *Cilantro (coriander): Benefits, nutrition, and preparation tips*. [online]. 2019 [cited 2021, April 15] Available from: URL: https://www.medicalnewstoday.com/articles/277627.

Wei, A., and T. Shibamoto, 2007. "Antioxidant activities and volatile constituents of various essential oils."*Journal of Agricultural and Food Chemistry*. 55:1737–1742.

Winarti, S., C. N. Pertiwi, A. Z. Hanani, S. I. Mujamil, K. A. Putra, and K. C. Herlambang, 2018. "Beneficial of coriander leaves (*Coriandrum sativum* L.) to reduce heavy metals contamination in rod shellfish." *Journal of Physics.* Conference Series.953:012237.

Yepez, B., Espinosa, M. S. López, and G. Bolaños, 2002. "Producing antioxidant fractions from herbaceous matrices by supercritical fluid extraction." *Fluid Phase Equilibria.* 194-197:879–884.

Yogesh, P, K. T. Luxmi, and P. Alok. *Cultivation of coriander* [online].2017 [cited 2021 April 12] Available from:https://www.krishisandesh.com/cultivation-of-coriander/.

Zare-Shehneh, M., M. Askarfarashah, and L. Ebrahimi, 2014. "Biological activities of a new antimicrobial peptide from *Coriandrum sativum.*" *International Journal of Biosciences*.6:89-99.

Zoubiri, S., and A. Baaliouamer, 2010. "Essential oil composition of *Coriandrum sativum* seed cultivated in Algeria as food grains protectant." *Food Chemistry*. 122:1226–1228.

Chapter 5

Importance of Coriander in the Food Industry

Dhriti Sharma[1,2], Savita Bhardwaj[1], Tunisha Verma[1], Mamta Pujari[1], Dhriti Kapoor[1] and Bharat Kapoor[3,*]

[1]Department of Botany, School of Bioengineering and Biosciences, Lovely Professional University, Phagwara (Punjab), India

[2]Sidharth Govt. College, Nadaun, Hamirpur (Himachal Pradesh), India

[3]Department of Hotel Management and Tourism, Guru Nanak Dev University, Amritsar (Punjab), India

Abstract

Coriander (*Coriandrum sativum* L.), an important herbaceous member of family *Apiaceae*, has been used for the purpose of seasoning since ages. All the parts of this plant as the likes of leaves, seeds/fruits and roots differ distinctly in their flavour from each other and are used variously to enhance the palatability of diverse food preparations. Citrusy overtones of its leaves confer them with fresh flavour and fragrance whereas the pungency of ripe seeds make them a class apart among all the spices. Entire plant is edible but quick to perish, so its shelf life can be prolonged by resorting to some of the best post-harvest management practices and processing techniques. A number of vital nutritional components constitute the inner build of this plant out of which essential oils (mainly linalool and furanocoumarins), polyphenols, fats like petroselenic acid and vitamins makes its consumption quite beneficial for human health. It means that this seed spice not only forms the irreplaceable part of various culinary items on a global level but also employed in different formulations in traditional and modern systems of medicines. Additionally, the essential oil obtained from this plant possesses antimicrobial and anti-oxidative properties which help in

* Corresponding Author's Email: bharatkapoor22@gmail.com.

In: *Coriandrum sativum*
Editors: Mamta Pujari, Bharat Kapoor, Neeta Pande et al.
ISBN: 978-1-68507-842-3

preserving the food preparations for longer time periods. Keeping all the nutritional and functional attributes of this plant in consideration, the aim of present study is to highlight its importance in food industry especially in terms of it being a flavouring agent, preservative and other value added products.

Keywords: coriander, seed spice, seasoning, food industry, processing, essential oil, antimicrobial, antioxidant, preservative

1. Introduction

Since historical times, spices being aromatic vegetable materials have been used extensively to season and flavour a variety of food preparations. Coriander (*Coriandrum sativum* L.; Vern. Dhania), a native of Mediterranean region, is one such common flavouring agent belonging to family *Apiaceae which got its cultivation dissipated to China, India and other parts of the world as an aromatic herb (Mandal and Mandal 2015).* Either the entire plant of this herbaceous spice or simply the individual parts like leaves, roots and seeds have been put to use in fresh as well as processed form. Leaves of this plant called as Cilantro or Chinese parsley are popular as the seasoning herb; the round seeds on account of their unique flavour and floral aroma are utilized in making spice or spice blends whereas its roots find a special inclusion in Thai curries for adding a peculiar pungent taste to them.

However, almost all the cuisines throughout the globe especially Asian, Latin and European, prefer the use of its fresh leaves and ripened fruits; though, both of these parts have been found to be different from each other in their fragrance and flavour. Additionally, its leaves and seeds yield an essential oil which deserve a significant mention because of its flavouring applicability in food products as the likes of beverages, sweets, pickles, fish and other meat items besides being used as a preservative. Moreover, this plant is not just an aroma additive, its medicinal worth because of its anti-cancerous, anti-oxidative, anti-inflammatory, antiseptic, carminative, diuretic, hepato-protective, hypoglycemic, hypocholesterolemic and stimulative potential, makes its incorporation in our dietary habits almost a necessity. Likewise, home remedies centred around the different parts of this plant are also present there which improves the overall health of our digestive system, further enhancing the importance of this spice crop in food industry.

Considering the importance of coriander plant as discussed above, its commercial cultivation is quite common all over the world with India exceling at all the levels, be it production, consumption or exportation of its seeds. Indian states namely Andhra Pradesh, Karnataka, Madhya Pradesh, Rajasthan, Tamil Nadu and Uttar Pradesh are leading growers of this herbaceous plant. Besides India, commercial cultivation of Coriander is prevalent in countries like Argentina, Australia, Morocco, Bulgaria, Canada, France, Italy, Romania, the Netherlands, Mexico, Myanmar, Pakistan, Spain, Turkey, and to some extent, in the Britain and America. India also tops as the world's largest exporter of coriander seeds, with 34% of the world's exports in 2019 followed by Italy (12%) and Russia (11.5%) (Trend Economy 2021).

In spite of having all the promising attributes of being a seasoning and flavouring spice along with that of a curative herb which tends to possess a vast potential of getting utilized variously in food and pharmaceutical industries, this crop still falls into the category of underused and overlooked plants (Beemnet and Getinet 2010; Kalidasu et al. 2015). Therefore, this chapter intends to furnish the details about the importance of this spice crop in terms of its usage, historical background, nutritional composition and applications in food industry in particular, so that its full potential could be effectively actualized in the coming future.

1.1. The World of Coriander with Myriad of Benefits at a Quick Glance

It is a well-known fact that for enhancing the flavour quotient, palatability and pungency of different food preparations, spices like coriander are added. Apart from this, coriander is also one of the popular domestic remedies against a whole array of digestion related problems as the likes of gastritis, food poisoning, loss of appetite, diarrhoea, hernia, colon spasms, piles and ulcers. Other health afflictions such as anaemia, inflammation of skin and conjunctiva, microbial infections, memory loss, vertigo, weakness, high cholesterol levels, measles, arthritis, toothache are also dealt with this plant quite effectively as mentioned and practised in modern and traditional systems of medicines. Miscellaneous other benefits also come along with this popular seed spice crop and all of its advantageous uses have been tabularized in brief as below in Table 1.

Table 1. Applicability of coriander in summarised form

<table>
<tr><th colspan="2">Different uses of Coriander</th><th>References</th></tr>
<tr><td>1. Flavouring agent</td><td>• As spice: a) in daily meal preparations all over the globe b) in soups like lemon coriander (Indian cuisine), Chinese cilantro (Chinese cuisine), Tom yam (Thai cuisine), Bun mam, Pho ga, Banh canh, Canh chua, Bun bo hue (Vietanamese cuisine).
• As puree and pastes in fast food industries</td><td>Diederichsen (1996); Mengesha et al. (2010); Geremew et al. (2015)</td></tr>
<tr><td rowspan="4">2. Healing herb</td><td>Ayurvedic: as rochan (laxative), deepan (appetizer), tridosh nashak (maintains balance between three doshas-vat, pitt and kaf), krimidhan (parasiticide), sheet prashaman (Cold curer), shool har (Pain killer)</td><td rowspan="4">Delaquis et al. (2002); Kubo et al. (2004); Nadeem et al. (2013); Singletary (2016)</td></tr>
<tr><td>Unani: For treating Nafkh-e-Shikam (flatulence), Suda (headache), Zof-e-Dimagh (weakness of brain), Zof-e-Qalb (cardiac weakness), and Zof-e-Meda (weakness of stomach)</td></tr>
<tr><td>Siddha: cures Navartci (dryness of mouth), Takam (thirst), Pun (ulcers), Carayaveri (alcoholic delirium), Vanti (emesis), Veppam (fever) and Kulirkaiccal (fever with rigor).</td></tr>
<tr><td>European: oral consumption in the form of herbal tea, fluid extract or essential oil to cure digestive disturbances and diabetes</td></tr>
<tr><td>3. Miscellaneous uses</td><td>• As a preservative for different food items
• As an ingredient in perfumes, cosmetics, soaps, detergents and other toiletries
• In Aromatherapy
• In Oleo-chemical industry
• As Plastic lubricant
• In the manufacturing of nylons
• In making oil cakes for ruminants
• As good melliferous agent, so cultivated for rearing honey bees</td><td>Bhat et al. (2014); Bhalchandra et al. (2014); Abou-Shaara (2015); Pacheco et al. (2016); Priyadarshi et al. (2016); Prachayasittikul et al. (2018)</td></tr>
</table>

2. Importance of Coriander in Food Industry in Terms of

2.1. Historical Journey so Far…

This plant has its cultivation going on since aeons and is ranked amongst the primarily used spices (Meena et al. 2014). However, its historical background can be unearthed back to at least 5000 BC. References to coriander can be found in Sanskrit writings and biblical testaments. In bible, it is mentioned that the "manna" offered to the Jews for fleeing Egypt resembled the coriander seeds. Owing to its stimulative properties, Egyptians referred to this plant as the "spice of happiness" and they used to burry coriander seeds in the tombs of their deceased rulers. Hanging gardens of Babylon, one of the seven wonders of the world, was reported to be fragrated by this herb. Romans and Greeks profusely used this plant; the former especially in the preservation of meats and flavoring of breads while the latter utilized the plant in perfumery, culinary items and as medicinal herb (Burdock and Carabin 2009; Abascal and Yarnell 2012). Romans were instrumental in introducing this herb into Britain where it gained popularity in English cuisine till the time the Renaissance happened, afterwards some other exotic spices took over. Coriander also finds its place as one of the first grown herbs by the American colonists of Massachusetts and 17th century French people were reported to make liquor from the distillation of this plant. Russia was the first country to establish steam distillation factory for the production of essential oil from coriander in 1885 (Eikani et al. 2007).

In Indian context, its fresh leaves were the most used part since the Vedic times whereas seeds got their due popularity as a spice after the arrival of Muslims whose Mughlai food items are a proof to it. Coming to the present times, coriander enjoys worldwide application being cultivated in both tropical and subtropical countries where this herb is incorporated as an indispensable part of their culinary traditions.

2.2. Nutritional Makeup

Phytonutrients (borneol, carvone, camphor, geraniol, elemol, limonene and linalool); flavonoids (epigenin, kaempferol, quercetin and rhamnetin); active phenolic compounds (caffeic and chlorogenic acid); vitamins (Vit. A, Vit. B_1, B_2, B_3, B_9, B_{12}, Vit. C, Vit. D and Vit. E); minerals (Calcium, Phosphorus, Iron, Manganese and Magnesium) and dietary fibres are present in different parts of coriander being responsible for all the flavouring plus healing attributes of this herb (Nimish et al. 2011; USDA 2016). For instance, fruits of coriander are rich in essential oils, fatty acids, alkaloids, anthraquinones, flavones, resins, sugars, sterols, tannins and fixed oils (Barros et al. 2012; Uitterhaegen et al. 2016). Monoterpenoid rich essential oil obtained after applying methods such as hydrodistillation, soxhlet extraction or simply supercritical water extraction show linalool as the main constituent (83%) followed by alpha-pinene, camphor, cis-dihydrocarvone, gamma-terpinene, geraniol, geranyl acetate and limonene, all making up only 10% of the total essential oil content (Eikani et al. 2007). The essential oil content has been reported to be much higher in small sized fruits of Caucasian and Central Asian varieties of this spice crop.

Out of all the fatty acids in fruits, petroselinic acid (cis-6-octadecenoic acid) is the principal component imparting unique physicochemical properties followed by linoleic acid, myristic acid, palmitic acid, stearic acid, oleic acid and vaccenic acid (Mandal and Mandal 2015; Beyzi et al. 2017). Oil cakes left after oil extraction are quite rich in carbohydrates like cellulose, proteins, fats, some nitrogen free extracts and ash which could be employed as animal feed (Ramadan and Morsel 2002; Kadiri et al. 2017). Water content of unripened coriander seeds has been reported to be 84% (Bhat et al. 2014). The plant has been reported to undergo a drastic change on its way to maturity, this explains why fruits/seeds are quite different from the herb with respect to their aroma. The mature fruits are with pleasant citrusy fragrance whereas the immature fruits along with leaves due to the presence of trans-tridecen component smell like a stink bug (Rajeshwari and Andallu 2011).

2.3. Post-Harvest Management and Processing

The innate genetic constitution of coriander cultivar and the timings of its harvest, play key roles in determining the size, volatile oil content, fragrance/flavour and quality of dried, mature fruit produce (Douglas et al. 2005). For the purpose of better marketing, removal of extraneous debris like stalks, plant refuse and soil from the fruits is a must. Recommended quality level is maintained by cleaning the product with the help of a de-stoner spiral gravity separator or vacuum gravity separators followed by grading plus packing in lint-free sacks and finally storing in non-humid, cool, pest free, well-aerated storehouses. Since storage is critical to the lifespan of a spice, all the storage factors are needed to be tended to carefully. Owing to a number of disadvantages of having this seed spice in raw form such as its bulkiness which makes storage difficult; its vulnerability for getting contaminated by microbes (bacteria and fungi); comparatively limited flavour and moreover, inconsistency of prices, turn the conversion of this crop commodity into value added products like ground spice, spice blends/ curry powders, oleoresins, volatile oil and consumer packed forms, all the more a favoured option at the global market level. All such alternatives are briefly discussed as follows:

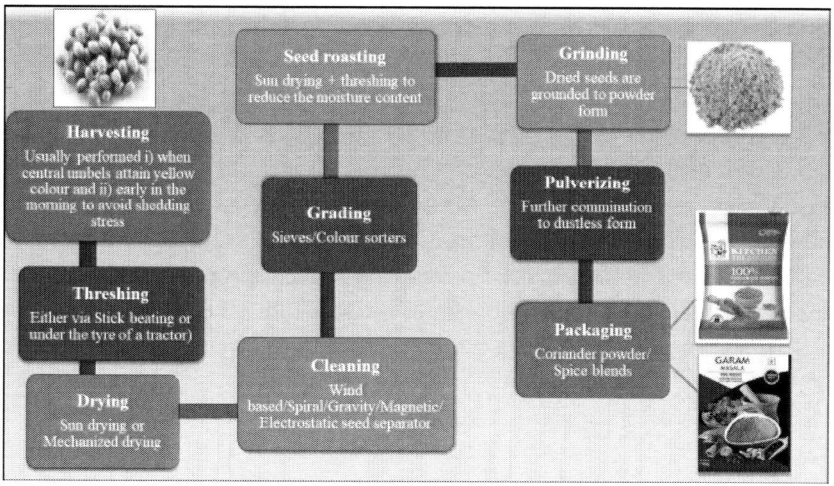

Figure 1. Flow diagram of processing of coriander seeds into consumer- packed curry powders or spice blends.

2.3.1. Ground Spices

Whole seeds having otherwise a coarse texture to chew, are subjected to milling to obtain desirable extents of fineness, usually inside a food processer while avoiding unwanted changes in the innate taste and pungency. Use of this form in various food preparations can be carried out in comparatively easy and uniform manner. However, grounded form of this seed spice akin to other spices tend to lose flavour with time, get contaminated with microbes or simply undergoes oxidation and all of this largely contribute to delimit its shelf life.

2.3.2. Consumer-Packed Spices

Both the whole or grounded forms of this seed spice are packaged variously keeping in mind the preferences of consumers belonging to three major sectors i.e., retail, industrial and institutional. Packaging has been found to prolong the period of sustenance of all the spices, especially inside the all new, improved and better packing materials as the likes of aluminium foils, laminations and plastic films by using advanced film-sealing machines. Choosing a particular style of packing helps in categorising the spice into paste/powder form or simply an instant one which results in fetching of more income per unit, even for the same quantity of bulk spices. The retail prices of spice packs stand at 50 to 100% higher vis-à-vis bulk spices. Though, the spice market is quite vast yet building of brand image for your packed spice has become a prerequisite for outsmarting the competition. The entire processing technique of conversion of coriander seeds into grounded form to be further utilized in curry powders and spice blends has been summarised in Figure 1.

2.3.3. Volatile Oil

The ripened seeds of this plant upon drying act as a major source of the volatile oil with a strong and distinct aroma. It is few of the most widely used essential oils which are known for their peculiar biological properties of being anti-oxidative, anti-diabetic, antidepressant, anti-epileptic, anti- hypertensive, anti-inflammatory, antimicrobial, anti-mutagenic, diuretic and neuroprotective by nature (Matasyoh et al. 2009; Sahib et al. 2012). The process of production of this oil involves placing the dried seeds in specifically designed distillation vessels of stainless steel furnished with an inlet for steam, outlet for the vapours and a separate set up for condensation and separation. Just below the charge, introduction of live steam is made, which when rises through the plant charge, yields volatile oil followed by its condensation and separation from water. The value added worth of this essential oil, if compared to other processed products has been reported to be low because of its presence in minimal quantities.

So, for obtaining this oil in certainly better amounts, super critical CO_2 extraction method with either CO_2 or propane as solvents, pressure at 200-300 bars and temperature at 35°C is employed more preferably. However, propane and propane containing liquids have comparatively greater solvating potential than CO_2 and can bring about an optimized oil recovery at a temperature of 25°C in pressure range of 50-100 bars (Illes et al. 2000). Obtained by any of these methods, the beneficial aspects of this oil involve the uniformity in the quality of its flavour, absence of enzymes and tannins along with non-hampering of the colour of final product. Moreover, the warm aromatic taste and typical linalool rich fragrance of this essential oil has been declared safe for human consumption by Food and Drug Administration (FDA), Council of Europe along with Foreign Exchange Management Act (FEMA) (Silva et al. 2011b).

2.3.4. Oleoresin

The non-volatile part of resinous nature having a flavour of its own is called as oleoresin and it is constituted mainly of pigments, antioxidants, fixative and heat components. Real essence of this seed crop is said to lie in its oleoresin part. Oleoresins have a free-flowing nature with 95% solubility in alcohol and colour wise, they vary from brown to yellow at room temperature. For obtaining this product, either grounded or roasted seeds are subjected to solvent extraction to get fruit like aroma with a slight balsamic tinge and caramel like taste respectively. Like that of essential oils, oleoresins are also quite low in their quantities, so not considered among value added products of prominence.

2.3.5. Pastes and Purees

Green leaves of this plant are pounded to make pastes and purees which in recent times have gained extensive popularity in fast food industries. Once the plant is harvested, its fresh parts tend to degrade which can be prevented via their conversion into valuable products as the likes of pastes, powders, sauces, salsas, purees, etc. For preparing the paste, blanching of leaves in steam (at 100°C) or water (90-100°C) for about ten minutes is preferably performed first, followed by their milling into paste. However, prior to transformation of both the leaves and fruits into paste, blanching in water is recommended for only a shorter time period (Kaiser et al. 2013). On the other hand, converting these leaves into powder form firstly requires blanching of fresh green leaves at 90°C for a minimum of 2 minutes so that action of peroxidase enzyme responsible for degradation of leaves is inhibited and then they are pulverized to uniform sized powder particles (Ahmed et al. 2004). The sauce prepared from coriander leaves is not spicy by nature, but it has deep green colour with unique and intense pungency.

2.4. Uses in Food Preparations

The common use of this spice crop is in flavouring various culinary preparations, for which its two principal components i.e., either the green herb or the dried seeds, are majorly employed. In young state, the entire plant is variously utilized, as in for making chutneys and green vegetable whereas the fresh leaves add a pleasant smell and unique flavour to food items such as curries, soups, sauces and a range of other dishes. Dried and grounded leaves can also be utilized by incorporating them into wheat flour for making breads with better anti-oxidative potential, moisture retention, shelf life, baking properties as well as sensory attributes with respect to aroma, texture and flavour. Optimum concentrations of coriander leaf powder being added to wheat flour for fortifying the breads lies in the range of 3.0 to 5.0% (w/w) (Das et al. 2012).

The whole seeds are an integral part of spice blends or curry powders along with other spices such as black pepper, chillies, cumin, fenugreek, fennel, onion and garlic, etc. Seeds when grounded coarsely have a crunchy texture and they form the bulkiest component of curry powders. This seed powder is used in brining, pickling and seasonings. Preparation of different food items ranging from buns, candies, cakes, pastries, stews and sausages too involve the use of its seeds. Even the alcoholic beverages and tobacco products are conferred with specific flavour of this seed spice. Some traditional sweetmeats and breath sweeteners can also be made from these seeds.

Likewise, oleoresin, being one of the value-added products of this spice, also finds its usage in seasoning different delicacies as the likes of pickles, sweets, beverages among numerous others. Additionally, this plant exhibits entomophily as its flowers are a rich source of nectar for attracting insect pollinators especially honey bees. Sieved coriander honey, thus produced, is popular for its unique taste and aromatic nature. Back in 1976, Lukjanav and Reznikov even reported that about 500 Kg of honey could be produced by honey bees from a coriander field measuring one hectare.

2.5. In Food Preservation

Food items undergo degradation on account of microbial action plus other physical and chemical changes and this has always been a matter of concern for the food industry. Moreover, recent years have witnessed growing awareness, pertinently at the levels of consumers, about possible health issues which may arise due to the use of synthetic preservatives. Considering the significance of these two issues i.e., i) food deterioration and ii) harmful impact of chemical/synthetic preservatives, more emphasis is now being laid to develop natural and safe alternatives to prolong the shelf life of foodstuffs. In this regard, essential oils are now finding their incorporation directly either into food items or indirectly in the packaging stuff owing to their potential of being used as a better preservative. Given their plant-based origin and long history of being utilized in cuisines and traditional medicinal systems, usage of essential oils in the form of food preservative stands quite undisputed. However, factors like degree of efficiency and approval of consumers are pre-requisites for their applicability in preservation. Further, the acceptance level of consumers depends upon the extent of modification in the sensory attributes of food item brought about by the addition of essential oil as well as the safety of consumption of the end product with regards to hypersensitivity or allergic responses (Michalczyk et al. 2012).

So far, a number of essential oils procured from plants such as marjoram (*Origanum marjorana*), oregano (*Origanum vulgare*), sage (*Salvia officinalis*) and thyme (*Thymus vulgaris*) have been utilized as preservative for prolonging the shelf life of fish and meat products on account of their bacteriostatic properties (Del Nobile et al. 2012). But if use of coriander oil in preserving food is considered in particular, an entire range of foodstuffs right from the alcoholic beverages, candy, chewing gum, dairy products, meat sausage, pickles to the tobacco products, is there which can be made to last long with this add-on in permissible limits of 0.1 to 100 ppm. Plentitude of pathogens such as *Campylobacter jejuni*, *Staphylococcus aureus* or *Escherichia coli*, *Listeria monocytogenes*, *Salmonella typhimurium*, *Yersinia enterocolitica* which spoil food by producing toxins of shiga and non-shiga nature have been found to be effectively dealt by adding the coriander oil plainly or incorporating the oil into packing films. In 2012, Michalczyk et al. reported a drastic reduction in bacterial counts particularly belonging to the *Enterobacteriaceae* family upon direct addition of 0.02% v/w coriander oil to minced beef, without unnecessarily changing the sensory nature of this delicacy.

On the similar lines, lower concentrations of coriander oil as 0.1 and 0.25% when added to beef and minced chicken meat bring down the bacterial loads whilst higher concentrations as 0.5% v/w cause cell death of *C. jejuni* after having a contact time of half an hour (Rattanachaikunsopon and Phumkhachorn 2010). However, Zivanovic et al. 2005 observed a significant reduction in the degree of efficacy of coriander oil vis-à-vis other oils against *E. coli* which produces shiga toxin and *L. monocytogenes* when infused into chitosan films. So far, the approach which proved successful in preserving food by using linalool component of essential oil to reduce the microbial contamination in cheddar cheese is when oil gets incorporated into low density polystyrene films (LDPFs) (Suppakul et al. 2008). These films yielded similar results when cheese was subjected to artificial contamination with *Listeria innocua* or *E. coli*. Their inhibitory action on the growth of *E. coli* remained effective despite being stored for periods as long as one year (Suppakul et al. 2011). Additionally, the linalool, given the percentage at which it is incorporated, does not cause any unnecessary changes in the gustatory properties of any food item.

To prove the efficacy of coriander oil against a number of microbes (bacterial and fungal species), different antimicrobial susceptibility trials such as diffusion (well or disk), agar dilution/broth dilution (macro and micro) have been adopted, the results of which are discussed in detail as follows:

2.5.1. Antibacterial Activity

Essential oil of coriander has been reported to be effectively operative against an extensive range of bacteria (Gram-positive and Gram-negative both) which normally devour upon foodstuffs to degrade them. Effective quantities of coriander oil capable of inhibiting growth of Shiga and non-Shiga toxin producing bacteria which includes a number of *Pseudomonas* species, *Bacillus megaterium*, *E. coli*, *L. monocytogenes*, *S. typhimurium*, *Y. enterocolitica*, were found to be 24 mg to 0.09 μg when loaded onto the paper disks during disk diffusion assays. Further, it has been reported that coriander oil has efficacy only towards gram-ve bacteria like *C. jejuni* and caused no harm to gram+ve bacteria such as *Bacillus cereus*, *Enterococcus faecalis* and *Lactobacillus plantarum* (Elgayyar et al. 2001; Rattanachaikunsopon and Phumkhachorn 2010; Silva et al. 2011b) whereas stark contrasting results were obtained by some of the research works where activity of coriander oil was observed to be more towards gram+ve bacteria like *L. monocytogenes* and *S. aureus* than the gram-ve ones as *Pseudomonas fragi* or *S. typhimurium* (Delaquis et al. 2002; Lo Cantore et al. 2004). Values like minimum inhibitory concentrations (MIC) and minimum bacterial concentrations (MBC) for evaluating the bacterial susceptibility can be tested by employing both agar and broth dilution techniques which yielded results on the similar lines with disk diffusion assays. It means efficacy of coriander oil against *Acinetobacter baumannii*, *C. jejuni*, *E. coli* (Both shiga and non-shiga toxin producing), *Enterococcus* sp. (vanomycin resistant), *Klebsiella pneumoniae*, *S. aureus* (both methillin-sensitive and methicillin-resistant ones) among all the other pathogens, has been proven. Further, it is widely accepted that the antimicrobial activity of essential oils depends on major constituents and their concentrations.

However, this bactericidal efficiency of coriander oil can be attributed to not only the major constituents but also the minor ones, though phenolic compounds top this antimicrobial efficacy among all the others such as aldehydes, alcohols, ethers, hydrocarbons and ketones (Ferdes and Ungureanu 2012). To sum up, it is actually the individual components of coriander oil as the likes of linalool, α-pinene, limonene, linalyl acetate, p-cymene and γ-terpinene which effectively fight against a range of a number of Gram-positive and Gram-negative bacteria (Trombetta et al. 2005; Di Pasqua et al. 2006; Sonboli et al. 2006; Cristani et al. 2007; Ozek et al. 2010). Theoverall antimicrobial efficiency of this essential oil is slightly higher than its diverse individual constituents owing to production of a cumulative effect on account of intricate interplay among them all.

2.5.2. Antifungal Activity

Agar/broth dilution as well as disk diffusion assays have proven antifungal nature of coriander essential oil where its application effectively fights out fungi like unicellular yeasts, certain hyphal forms along with skin infection causing dermatophytes. Further, most of the studies regarding antifungal efficiency of coriander oil revolve around *Candida* sp., apparently because of its medical significance (Furletti et al. 2011), though the oil is also found to be capable of inhibiting growth of other fungi as the likes of *Aspergillus niger*, *Geotrichum candidum*, *Kluyveromyces fragilis*, *Microsporum canis* and *Saccharomyces cerevisiae* (Delaquis et al. 2002; Toroglu 2011). Disk diffusion assays helped in working out concentrations of coriander oil effective for hampering the growth of *Candida* sp. which came out to be in the range of 0.0008% to 0.4% whereas for *M. canis* and *S. cerevisiae*, these were 0.009-0.07% and 0.13% respectively. The minimum lethal concentrations (MLC) value for assessing the fungicidal potential was also determined in some of these research works which ranged from 0.017 to 0.4% (Silva et al. 2011a; Soares et al. 2012).

2.5.3. Other Antimicrobial Activities

Through all these years, the use of antimicrobial agents is on the rise which has led to the development of several antibiotic resistant microbial (bacterial and fungal) varieties popularly called as superbugs (Teuber 1999). To win the battle against them, more concerted efforts are being put by the researchers by employing diverse combinations of compounds of natural origin which are in turn expected to have the antimicrobial potential or capacity to reverse the phenomenon of antibiotic resistance. Some workers while experimenting along these lines assessed the synergistic effect of coriander oil in response to various antifungal and antibiotic preparations. They have reported its additive efficacy with antibiotics ceftriaxone for *M. smegmatis*, ceftriaxone + gentamacin for *S. aureus* and ciprofloxacin + tetracycline + gentamicin + chloroamphenicol against *A. baumannii*, though some of the results revealed antagonism among coriander oil and certain antibiotics against bacteria of both gram +ve and –ve nature (Toroglu 2011; Duarte et al. 2012).

Furthermore, coriander oil has also been found to reduce the formation of microbial biofilms which could otherwise enhance the microbial resistance against antibiotics and other disinfectants. This anti-biofilm activity of coriander oil has been mostly studied in case of planktonic cells and further, a concentration of 0.125 mg/mL was found to be potent in hampering the formation of biofilms by *Acinetobacter baumannii* and *Candida albicans* via lag phase enhancement (Furletti et al. 2011). Coriander oil has been reported to prevent formation of biofilms in concentrations double to quadruple times the MIC value (0.2-1.6%), which in turn brings about an eighty-five percent decrease in total biomass and metabolic activity of biofilms. Same concentrations of this oil within 24 hours of incubation cause a reduction of 75-90% in total biomass and metabolic efficiency of pre-formed biofilms (Duarte et al. 2013).

These results make the prospects of coriander oil for being used as an anti-biofilm agent quite promising because efficacy of antibiotics against cells found in a biofilm is reported to be thousand times lower when compared to planktonic cells (Melchior et al. 2006). Another important feature of essential oils is their volatility but currently there is a general lack of extensive scientific information about the effectiveness of essential oils in the vapour phase. However, its evaluation has been gaining interest in the recent years. For instance, vapourised form of linalool component of coriander oil has been reported to prevent the growth of *C. albicans* and *S. choleraesuis* more effectively as compared to its aqueous form owing to the fact that contact of essential oil with microbes becomes easier in vapour phase whereas in aqueous phase, lipophilic components of oil form micelles preventing their association with microorganisms (Lopez et al. 2005; Lopez et al. 2007b; Martinez-Abad et al. 2013).

2.5.3.1. Mechanism of Action behind Antimicrobial Potential

In spite of numerous extensive research works, the exact course of action lying behind antimicrobial nature of coriander oil is yet to be completely deciphered. Unveiling the mode of action of essential oils on bacterial and fungal cells is of great relevance towards their successful application in the pharmaceutical and food industries. So far, the broad mode of action of coriander oil against bacteria of both gram +ve and –ve nature has been found to be the same; however bacterial sensitivity differs because of variability in organisation of cell wall in both the bacterial types and the composition of oil employed for inhibition. For instance, in case of gram+ve bacteria, the coriander oil application at low concentrations makes the cell wall become thicker, probably owing to presence of alcohols in the oil, especially linalool whereas in Gram-negative bacteria, (+)-linalool strongly interacts with outer lipopolysaccharide layer of cell wall to bring about disruption of its structure to make entry of oil into the intercellular space easy (Linhova et al. 2010).

Several other studies also substantiate this action of oil on bacterial cell which further go on to reveal that the other components of coriander oil like linalyl acetate, limonene, γ-terpinene and p-cymene bind with phospholipids of cell membranes in both gram+ve and –ve bacteria to cause homeoviscous changes and leakage in bacterial cell by disrupting the lipid composition of cell membranes (Trombetta et al. 2005; Di Pasqua et al. 2006; Cristani et al. 2007). Within almost half an hour of incubation, this oil kills bacteria by enhancing permeability and efflux activity of membranes leading to inhibition of vital processes of cellular metabolism (Silva et al. 2011b). On the other hand, the antifungal nature of this oil works via prevention of formation of germ tube in fungus *Candida albicans*, even if applied at concentrations lower than MIC measures. Besides this, coriander oil on the similar lines with its bacterial action, damage the fungal membranes of all cells, thereby leading to an increase in membrane permeability followed by leakage of intracellular parts even as large as that of DNA within half an hour of incubation time (Silva et al. 2011a). All these mechanisms deciphered so far with regard to antibacterial and antifungal action of coriander oil are summarised in the Figure 2.

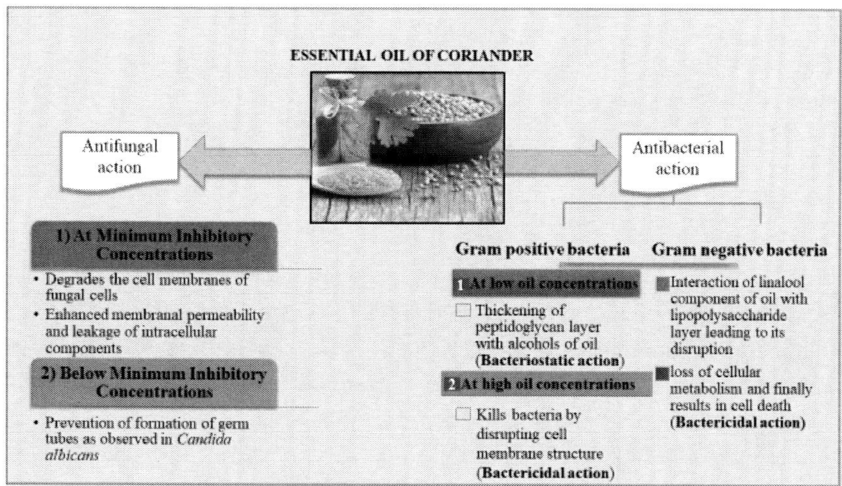

Figure 2. Mechanism of antimicrobial action of essential oil of coriander against foodborne pathogens (Modified after Silva and Domingues 2017).

2.5.4. Anti-Oxidative Potential of Coriander Essential Oil

Apart from getting degraded by microbes, food preparations also undergo changes in the nature of their chemical constituents such as fats, pigments, etc. mostly via oxidative methods. This oxidative deterioration or spoilage of food items can be effectively prevented by the addition of essential oils (Nerin et al. 2006; Bentayeb et al. 2007; Lopez de Dicastillo et al. 2011). Essential oil of coriander being no exception to it, has proven its antioxidative efficiency via enhancing the oxidative stability of the food items, as specifically noted in case of Italian salami, where the results in terms of prolonging the shelf life and maintaining the sensory profile are far better in comparison to a synthetic preservative (Marangoni and de Moura 2011a; Marangoni and de Moura 2011b). One another report further substantiated this anti-oxidative potential of coriander oil, where its application prevented the oxidative degradation of clarified butter or ghee, however, the effectiveness of this oil turned out to be maximum during the process of frying as compared to storage with antioxidants of synthetic nature (Patel et al. 2013)

This anti-oxidative activity of essential oil revolves around scavenging of free radicals, chelation of ions of transition metals or stoppage of lipid peroxidation as exhibited by various analytical studies, though in case of coriander oil, it is strongly correlated to either the presence of phenolic compounds or flavonoids and tannins (Samojlik et al. 2010; Sriti et al. 2011). However, care has to be taken while using coriander oil as an antioxidant on account of some reports hinting at its slight prooxidant effect on lipid peroxidation (Singh et al. 2002).

2.6. Safety and Substitution Issues

Fruits or seeds of coriander have been reported to cause no mortality in case of rats at a dose of 750mg/Kg, though higher doses might lead to infertility in female rats. Plant extract or juice was found to be non-toxic as well as non-mutagenic in the concentration range of 50-1000 μL/coincubation flask with a chlorophyll content of 0.0325 μg/mL (Cortes-Eslava et al. 2004). Further, the essential oil of coriander is tested for its safety in terms of its partial lethal dose i.e., (LD50) and it came out to be 2.257 ml/kg hinting at its low toxicity as per the values worked out by Ozbek et al. in 2006. It was further substantiated with pharmaco-genomic analysis where Coriander leaves Essential Oil (CLEO) along with its selected active components exhibited low cytotoxicity in case of human cells, thereby making the oil safe for consumption, mainly as a food preservative (Freires et al. 2014).

Besides this, sometimes the alternative members from the family *Apiaceae* such as sawtooth coriander/Mexican coriander (*Eryngium foetidum*) or from the family *Polygonaceae* like Vietnamese coriander *Persicaria odorata* are used as substitutes for getting aromatic flavour of coriander. Further, production of the essential oil of coriander at commercial level often come across its adulteration with anise fruit/anethole oil, cedar wood oil, sweet orange oil and turpentine oil which can also pose serious problems if consumed.

Conclusion

Coriander is an amazing plant among a wide array of herbs and spices which is added into foodstuffs since time immemorial for the purpose of seasoning along with rendering diverse health benefits. The multifariously beneficial nature in terms of being nutritive, protective and prophylactic herb has made it quite aptly popular as the "herb of happiness". It can grow round the year, but preferred to be processed for enhancing savouriness of different delicacies as well as its shelf life for domestic and international trade. Its fruits/seeds and leaves are strongly aromatic plus nutritionally rich in proteins, fats, minerals, vitamins, etc. and therefore multiple advantages with regard to human health are conferred with its inclusion in our daily dietary routines. The principle and well characterized attributes of this plant's essential oil as the likes of its anti-oxidative nature, broad spectrum antimicrobial efficacy along with ability to kill foodborne pathogens within merely half an hour of incubation make it an appealing and valuable choice for the preservation of various food preparations. Further, the application of essential oil of coriander i) either in combination with other preservatives or ii) simply by internalising it inside the packaging films along with iii) synergistically using it in advanced preservation strategies such as irradiation or modified atmosphere packaging (MAP), have been reported to be efficient enough to overcome the organoleptical changes, otherwise brought about by the essential oils in the food items. Coriander oil is ranked second at the global level with regards to its consumption, however, extensive clinical and research studies are still needed to be carried out to firmly establish its *in vivo* effectiveness and complete safety for consumption in food industries.

References

Abascal, Kathy, and Eric Yarnell. (2012). "Cilantro—Culinary Herb or Miracle Medicinal Plant?" *Alternative and Complementary Therapies*, 18, no. 5: 259-264.

Abou-Shaara, H. F. 2015. "Potential Honey Bee Plants of Egypt." *Cercetari Agronomice in Moldova*, 48(2), 99–108.

Ahmed, Jasim, U. S. Shivhare, and P. Singh. 2004. "Colour kinetics and rheology of coriander leaf puree and storage characteristics of the paste." *Food Chemistry*, 84, no. 4: 605-611.

Barros, Lillian, Montserrat Duenas, Maria Inês Dias, Maria João Sousa, Celestino Santos-Buelga, and Isabel CFR Ferreira. (2012) "Phenolic profiles of *in vivo* and *in vitro* grown *Coriandrum sativum* L." *Food Chemistry*, 132, no. 2: 841-848.

Beemnet, M. and Alemaw, G. 2010. "Variability in Ethiopian Coriander Accessions for Agronomic and Quality Traits." *Afr. Crop Sci. J.*, 18(2): 43–49.

Bentayeb, K., C. Rubio, C. Sanchez, R. Batlle, and C. Nerin. 2007. "Determination of the shelf life of a new antioxidant active packaging." *Italian Journal Of Food Science*, 19 (2007): 110-115.

Beyzi, Erman, Kevser Karaman, Adem Gunes, and Selma Buyukkilic Beyzi. 2017. "Change in some biochemical and bioactive properties and essential oil composition of coriander seed (*Coriandrum sativum* L.) varieties from Turkey." *Industrial Crops and Products*, 109: 74-78.

Bhalchandra, Waykar, R. K. Baviskar, and T. B. Nikam. 2014. "Diversity of nectariferous and polleniferous bee flora at Anjaneri and Dugarwadi hills of Western Ghats of Nasik district (MS) India." *Journal of Entomology and Zoology Studies*, 2, no. 4: 244-249.

Bhat, S., P. Kaushal, M. Kaur, and H. K. Sharma. 2014. "Coriander (*Coriandrum sativum* L.): Processing, nutritional and functional aspects." *African Journal of plant science*, 8, no. 1: 25-33.

Burdock, George A., and Ioana G. Carabin. (2009). "Safety assessment of coriander (*Coriandrum sativum* L.) essential oil as a food ingredient." *Food and Chemical Toxicology*, 47, no. 1: 22-34.

Casetti, F., S. Bartelke, K. Biehler, M. Augustin, C. M. Schempp, and U. Frank. 2012. "Antimicrobial activity against bacteria with dermatological relevance and skin tolerance of the essential oil from *Coriandrum sativum* L. fruits." *Phytotherapy research*, 26, no. 3: 420-424.

Cortés-Eslava, Josefina, Sandra Gómez-Arroyo, Rafael Villalobos-Pietrini, and Jesús Javier Espinosa-Aguirre. 2004. "Antimutagenicity of coriander (*Coriandrum sativum*) juice on the mutagenesis produced by plant metabolites of aromatic amines." *Toxicology letters*, 153, no. 2: 283-292.

Cristani, Mariateresa, Manuela D'Arrigo, Giuseppina Mandalari, Francesco Castelli, Maria Grazia Sarpietro, Dorotea Micieli, Vincenza Venuti, Giuseppe Bisignano, Antonella Saija, and Domenico Trombetta. 2007. "Interaction of four monoterpenes contained in essential oils with model membranes: implications for their antibacterial activity." *Journal of agricultural and food chemistry*, 55, no. 15: 6300-6308.

Das, Lipi, Utpal Raychaudhuri, and Runu Chakraborty. 2012. "Supplementation of common white bread by coriander leaf powder." *Food Science and Biotechnology*, 21, no. 2: 425-433.

Del Nobile, Matteo Alessandro, Annalisa Lucera, Cristina Costa, and Amalia Conte. "Food applications of natural antimicrobial compounds." *Frontiers in microbiology*, 3 (2012): 287.

Delaquis, Pascal J., Kareen Stanich, Benoit Girard, and G. Mazza. 2002. "Antimicrobial activity of individual and mixed fractions of dill, cilantro, coriander and eucalyptus essential oils." *International journal of food microbiology*, 74, no. 1-2 (2002): 101-109

Di Pasqua, Rosangela, Nikki Hoskins, Gail Betts, and Gianluigi Mauriello. 2006. "Changes in membrane fatty acids composition of microbial cells induced by addiction of thymol, carvacrol, limonene, cinnamaldehyde, and eugenol in the growing media." *Journal of agricultural and food chemistry*, 54, no. 7: 2745-2749.

Diederichsen, A. 1996. "Coriander (*Coriandrum sativum* L.). promoting the conservation and use of underutilized and negleted crops. 3. Rome: Institute of Plant Genetics and Crop Plant Research, Gatersleben." *International Plant Genetic Resources Institute*, 83.

Douglas, M., J. Heyes, and B. Smallfield. 2005. "Herbs, spices and essential oils: post-harvest operations in developing countries." *UNIDO and FAO*, 61: 1-8.

Duarte, A., S. Ferreira, F. Silva, and F. C. Domingues. 2012. "Synergistic activity of coriander oil and conventional antibiotics against *Acinetobacter baumannii*." *Phytomedicine*, 19, no. 3-4: 236-238.

Duarte, Andreia F., Susana Ferreira, Rosário Oliveira, and Fernanda C. Domingues. 2013. "Effect of coriander oil (*Coriandrum sativum*) on planktonic and biofilm cells of *Acinetobacter baumannii*." *Natural Product Communications*. 8, no. 5: 1934578X1300800532.

Eikani, Mohammad H., Fereshteh Golmohammad, and Soosan Rowshanzamir. (2007). "Subcritical water extraction of essential oils from coriander seeds (*Coriandrum sativum* L.)." *Journal of Food Engineering*, 80, no. 2: 735-740.

Elgayyar, M., F. A. Draughon, D. A. Golden, and J. R. Mount. 2001. "Antimicrobial activity of essential oils from plants against selected pathogenic and saprophytic microorganisms." *Journal of food protection*, 64, no. 7: 1019-1024.

Ferdeş, Mariana, and Camelia Ungureanu. 2012. "Antimicrobial activity of essential oils against four food-borne fungal strains." *University Politehnica of Bucharest Scientific Bulletin*, 74, no. 2.

Freires, Irlan de Almeida, Ramiro Mendonça Murata, Vivian Fernandes Furletti, Adilson Sartoratto, Severino Matias de Alencar, Glyn Mara Figueira, Janaina Aparecida de Oliveira Rodrigues, Marta Cristina Teixeira Duarte, and Pedro Luiz Rosalen. "*Coriandrum sativum* L. (coriander) essential oil: antifungal activity and mode of action on Candida spp., and molecular targets affected in human whole-genome expression." *PLoS One*, 9, no. 6: e99086.

Furletti, V. F., I. P. Teixeira, G. Obando-Pereda, R. C. Mardegan, A. Sartoratto, G. M. Figueira, R. M. T. Duarte, V. L. G. Rehder, M. C. T. Duarte, and J. F. Höfling. 2011.

"Action of *Coriandrum sativum* L. essential oil upon oral *Candida albicans* biofilm formation." *Evidence-Based Complementary and Alternative Medicine*.

Geremew, Awas, Mekbib Firew, and Ayana Amsalu. (2015). "Variability, heritability and genetic advance for some yield and yield related traits and oil content in Ethiopian coriander (*Coriandrum sativum* L.) genotypes." *International Journal of Plant Breeding and Genetics*, 9, no. 3: 116-125.

Illés, V., H. G. Daood, S. Perneczki, L. Szokonya, and M. Then. 2000. "Extraction of coriander seed oil by CO_2 and propane at super-and subcritical conditions." *The Journal of Supercritical Fluids*, 17, no. 2: 177-186.

Kadiri, L., A. Lebkiri, E. H. Rifi, Y. Essaadaoui, A. Ouass, I. Lebkiri, and H. Hamad. 2017. "Characterization of coriander seeds "*Coriandrum sativum*"." *International Journal of Scientific and Engineering Research*, 8, no. 7: 2303-2308.

Kaiser, Andrea, Dietmar R. Kammerer, and Reinhold Carle. 2013. "Impact of blanching on polyphenol stability and antioxidant capacity of innovative coriander (*Coriandrum sativum* L.) pastes." *Food Chemistry*, 140, no. 1-2: 332-339.

Kalidasu, Giridhar, S. Suryakumari, C. Sarada, A. Rajani, N. Hariprasada Rao, S. R. Pandravada, and L. Naidu. (2015). "Exploiting genetic divergence for crop improvement in coriander (*Coriandrum sativum* L.): A neglected and underutilized crop." *Indian Journal of Plant Genetic Resources*, 28, no. 2: 222-228.

Kubo, Isao, Ken-ichi Fujita, Aya Kubo, Ken-ichi Nihei, and Tetsuya Ogura. 2004. "Antibacterial activity of coriander volatile compounds against Salmonella choleraesuis." *Journal of agricultural and food chemistry*, 52, no. 11: 3329-3332.

Linhová, M., P. Patáková, J. Lipovský, P. Fribert, L. Paulová, M. Rychtera, and K. Melzoch. 2010. "Development of flow cytometry technique for detection of thinning of peptidoglycan layer as a result of solvent production by *Clostridium pasteurianum*." *Folia microbiologica*, 55, no. 4: 340-344.

Lo Cantore, Pietro, Nicola S. Iacobellis, Adriana De Marco, Francesco Capasso, and Felice Senatore. 2004. "Antibacterial activity of *Coriandrum sativum* L. and Foeniculum vulgare Miller var. vulgare (Miller) essential oils." *Journal of agricultural and food chemistry*, 52, no. 26: 7862-7866.

López de Dicastillo, Carol, Cristina Nerín, Pilar Alfaro, Ramón Catalá, Rafael Gavara, and Pilar Hernández-Muñoz. 2011. "Development of new antioxidant active packaging films based on ethylene vinyl alcohol copolymer (EVOH) and green tea extract." *Journal of agricultural and food chemistry*, 59, no. 14: 7832-7840.

Lopez, P., C. Sanchez, R. Batlle, and C. Nerin. 2005. "Solid-and vapor-phase antimicrobial activities of six essential oils: susceptibility of selected foodborne bacterial and fungal strains." *Journal of agricultural and food chemistry*, 53, no. 17: 6939-6946.

López, Patricia, Cristina Sanchez, Ramón Batlle, and Cristina Nerín. 2007. "Vapor-phase activities of cinnamon, thyme, and oregano essential oils and key constituents against foodborne microorganisms." *Journal of agricultural and food chemistry*, 55, no. 11: 4348-4356.

Luk'janov, I.A. and A.R. Reznikov. 1976. *Koriandr* [*Coriander*]. In: Smoljanova, A.M. & Ksendza, A.T. (Editors): Efirnomaslicnye kul'tury. Kolos, Moskva, Russia. pp. 9-57.

Mandal, S. and Manisha, Mandal. 2015. "Coriander (*Coriandrum sativum* L.) essential oil: Chemistry and biological activity." *Asian Pacific Journal of Tropical Biomedicine*, 5, no. 6: 421-428.

Marangoni, Cristiane, and Neusa Fernandes de Moura. 2011a. "Antioxidant activity of essential oil from *Coriandrum sativum* L. in Italian salami." *Food Science and Technology*, 31: 124-128.

Marangoni, Cristiane, and Neusa Fernandes de Moura. 2011b. "Sensory profile of Italian salami with coriander (*Coriandrum sativum* L.) essential oil." *Food Science and Technology*, 31: 119-123.

Martínez-Abad, A., G. Sánchez, V. Fuster, J. M. Lagaron, and M. J. Ocio. 2013. "Antibacterial performance of solvent cast polycaprolactone (PCL) films containing essential oils." *Food Control*, 34, no. 1: 214-220.

Matasyoh, J. C., Z. C. Maiyo, R. M. Ngure, and R. Chepkorir. 2009. "Chemical composition and antimicrobial activity of the essential oil of *Coriandrum sativum*." *Food Chemistry*, 113, no. 2: 526-529.

Meena, S. K., N. L. Jat, B. Sharma, and V. S. Meena. 2014. "Effect of plant growth regulators and sulphur on productivity of coriander (*Coriandrum sativum* L.) in Rajasthan." *J. Environ. Sci. Int*, 6: 69-73.

Melchior, M. B., H. Vaarkamp, and J. Fink-Gremmels. 2006. "Biofilms: a role in recurrent mastitis infections?" *The Veterinary Journal*, 171, no. 3: 398-407.

Mengesha, Beemnet, Getinet Alemaw, and Bizuayhu Tesfaye. (2010). "Performance of Ethiopian coriander (*Coriandrum sativum* L.) accessions in vegetative, phenological, generative and chemical characters." *Improving quality production of horticultural crops for sustainable development proceedings, February*, 04-05.

Michalczyk, Magdalena, Ryszard Macura, Iwona Tesarowicz, and Joanna Banaś. 2012. "Effect of adding essential oils of coriander (*Coriandrum sativum* L.) and hyssop (*Hyssopus officinalis* L.) on the shelf life of ground beef." *Meat science*, 90, no. 3: 842-850.

Nadeem, Muhammad, Faqir Muhammad Anjum, Muhammad Issa Khan, Saima Tehseen, Ahmed El-Ghorab, and Javed Iqbal Sultan. (2013). "Nutritional and medicinal aspects of coriander (*Coriandrum sativum* L.): A review." *British Food Journal*.

Nerín, Cristina, Laura Tovar, Djamel Djenane, Javier Camo, Jesús Salafranca, José Antonio Beltrán, and Pedro Roncalés. 2006. "Stabilization of beef meat by a new active packaging containing natural antioxidants." *Journal of Agricultural and Food Chemistry*, 54, no. 20: 7840-7846.

Nimish, P. L., K. B. Sanjay, B. M. Nayna, R. D. Jaimik. 2011. "Phytopharmacological properties of *Coriander sativum* as a potential medicinal tree: An Overview." *J. Appl. Pharm.*, 01(04): 20–25.

Ozbek, Hanefi, A. Tas, F. Ozgokce, N. Selcuk, S. Alp, and S. Karagoz. 2006. "Evaluation of median lethal dose and analgesic activity of *Foeniculum vulgare* Miller essential oil." *Int J Pharmacol*, 2, no. 2: 181-3.

Özek, Gulmira, Fatih Demirci, Temel Özek, Nurhayat Tabanca, David E. Wedge, Shabana I. Khan, Kemal Hüsnü Can Başer, Ahmet Duran, and Ergin Hamzaoglu. 2010. "Gas chromatographic–mass spectrometric analysis of volatiles obtained by four different techniques from *Salvia rosifolia* Sm., and evaluation for biological activity." *Journal of Chromatography A*, 1217, no. 5: 741-748.

Pacheco L, Neith A., Julia Cano-Sosa, Fernando Poblano C, Ingrid M. Rodríguez-Buenfil, and Ana Ramos-Díaz. 2016. "Different responses of the quality parameters of *Coriandrum sativum* to organic substrate mixtures and fertilization." *Agronomy*, 6, no. 2: 21.

Patel, Shriyesh, Sneha Shende, Sumit Arora, and Ashish K. Singh. 2013. "An assessment of the antioxidant potential of coriander extracts in ghee when stored at high temperature and during deep fat frying." *International Journal of Dairy Technology*, 66, no. 2: 207-213.

Prachayasittikul, Veda, Supaluk Prachayasittikul, Somsak Ruchirawat, and Virapong Prachayasittikul. 2018. "Coriander (*Coriandrum sativum*): A promising functional food toward the well-being." *Food Research International*, 105 (2018): 305-323.

Priyadarshi, Siddharth, Hafeeza Khanum, Ramasamy Ravi, Babasaheb Baskarrao Borse, and Madeneni Madhava Naidu. 2016. "Flavour characterisation and free radical scavenging activity of coriander (*Coriandrum sativum* L.) foliage." *Journal of food science and technology*, 53, no. 3: 1670-1678.

Rajeshwari, Ullagaddi, and Bondada Andallu. 2011. "Medicinal benefits of coriander (*Coriandrum sativum* L)." *Spatula DD*, 1, no. 1: 51-58.

Ramadan, Mohamed, and Jörg-Thomas Mörsel. 2002. "Oil composition of coriander (*Coriandrum sativum* L.) fruit-seeds." *European Food Research and Technology*, 215, no. 3: 204-209.

Rattanachaikunsopon, Pongsak, and Parichat Phumkhachorn. 2010. "Potential of coriander (*Coriandrum sativum*) oil as a natural antimicrobial compound in controlling *Campylobacter jejuni* in raw meat." *Bioscience, biotechnology, and biochemistry*, 74, no. 1: 31-35.

Sahib, Najla Gooda, Farooq Anwar, Anwarul-Hassan Gilani, Azizah Abdul Hamid, Nazamid Saari, and Khalid M. Alkharfy. 2013. "Coriander (*Coriandrum sativum* L.): A potential source of high-value components for functional foods and nutraceuticals-A review." *Phytotherapy Research*, 27, no. 10: 1439-1456.

Samojlik, Isidora, Neda Lakic, Neda Mimica-Dukic, Kornelia Đaković-Svajcer, and Biljana Bozin. 2010. "Antioxidant and hepatoprotective potential of essential oils of coriander (*Coriandrum sativum* L.) and caraway (*Carum carvi* L.) (Apiaceae)." *Journal of Agricultural and Food Chemistry*, 58, no. 15: 8848-8853.

Silva, Filomena, and Fernanda C. Domingues. 2017. "Antimicrobial activity of coriander oil and its effectiveness as food preservative." *Critical reviews in food science and nutrition*, 57, no. 1: 35-47.

Silva, Filomena, Susana Ferreira, Andreia Duarte, Dina I. Mendonca, and Fernanda C. Domingues. 2011a. "Antifungal activity of *Coriandrum sativum* essential oil, its mode of action against Candida species and potential synergism with amphotericin B." *Phytomedicine*, 19, no. 1: 42-47.

Silva, Filomena, Susana Ferreira, João A. Queiroz, and Fernanda C. Domingues. 2011b."Coriander (*Coriandrum sativum* L.) essential oil: its antibacterial activity and mode of action evaluated by flow cytometry." *Journal of medical microbiology*, 60, no. 10: 1479-1486.

Singh, G., I. P. S. Kapoor, S. K. Pandey, U. K. Singh, and R. K. Singh. 2002. "Studies on essential oils: part 10; antibacterial activity of volatile oils of some spices." *Phytotherapy Research: An International Journal Devoted to Pharmacological and Toxicological Evaluation of Natural Product Derivatives*, 16, no. 7: 680-682.

Singletary, Keith. 2016. "Coriander: overview of potential health benefits." *Nutrition today*, 51, no. 3: 151-161.

Soares, Bruna V., Selene M. Morais, Raquel Oliveira dos Santos Fontenelle, Vanessa A. Queiroz, Nadja S. Vila-Nova, Christiana Pereira, Edy S. Brito, C.S. Cavalcante, D.S. Castelo-Branco. 2012. "Antifungal activity, toxicity and chemical composition of the essential oil of *Coriandrum sativum* L. fruits." *Molecules*, 17, no. 7: 8439-8448.

Sonboli, Ali, Babak Babakhani, and Ahmad Reza Mehrabian. 2006. "Antimicrobial activity of six constituents of essential oil from Salvia." *Zeitschrift für Naturforschung C*, 61, no. 3-4: 160-164.

Sriti, Jazia, Wissem Aidi Wannes, Thierry Talou, Gerard Vilarem, and Brahim Marzouk. 2011. "Chemical composition and antioxidant activities of Tunisian and Canadian coriander (*Coriandrum sativum* L.) fruit." *Journal of Essential Oil Research*, 23, no. 4: 7-15.

Suppakul, Panuwat, Kees Sonneveld, Stephen W. Bigger, and Joseph Miltz. 2008. "Efficacy of polyethylene-based antimicrobial films containing principal constituents of basil." *LWT-Food Science and Technology*, 41, no. 5: 779-788.

Suppakul, Panuwat, Kees Sonneveld, Stephen W. Bigger, and Joseph Miltz. 2011. "Loss of AM additives from antimicrobial films during storage." *Journal of Food Engineering*, 105, no. 2: 270-276.

Telci, Isa, Ozlem Gul Toncer, and Nermin Sahbaz. 2006. "Yield, essential oil content and composition of *Coriandrum sativum* varieties (var. vulgare Alef and var. microcarpum DC.) grown in two different locations." *Journal of Essential Oil Research*, 18, no. 2: 189-193.

Teuber, M. 1999. "Spread of antibiotic resistance with food-borne pathogens." *Cellular and Molecular Life Sciences CMLS*, 56, no. 9: 755-763.

Toroglu, Sevil. 2011. "*In-vitro* antimicrobial activity and synergistic/antagonistic effect of interactions between antibiotics and some spice essential oils." *Journal of Environmental Biology*, 32, no. 1: 23-29.

Trend Economy. World Merchandise Exports and Imports by Commodity (HS02): *Seeds of Coriander*. Sofia, Bulgaria, 2021.

Trombetta, Domenico, Francesco Castelli, Maria Grazia Sarpietro, Vincenza Venuti, Mariateresa Cristani, Claudia Daniele, Antonella Saija, Gabriela Mazzanti, and Giuseppe Bisignano. 2005. "Mechanisms of antibacterial action of three monoterpenes." *Antimicrobial agents and chemotherapy*, 49, no. 6: 2474-2478.

Uitterhaegen, Evelien, Klicia A. Sampaio, Elisabeth IP Delbeke, Wim De Greyt, Muriel Cerny, Philippe Evon, Othmane Merah, Thierry Talou, and Christian V. Stevens. (2016). "Characterization of French coriander oil as source of petroselinic acid." *Molecules*, 21, no. 9:1202.

USDA. 2016. *National nutrient database for standard reference* release 28 full report (all nutrients). Nutrient data for, Spices, coriander seed.

Zivanovic, Svetlana, Shuang Chi, and Ann F. Draughon. 2005. "Antimicrobial activity of chitosan films enriched with essential oils." *Journal of food science*, 70, no. 1: M45-M51.

Chapter 6

Therapeutic Uses of Coriander

Sadaf Jan[1], Rattandeep Singh[1], Renu Bhardwaj[2], Monika Thakur[3] and Vandana Gautam[4,*]

[1]Department of Biotechnology, School of Bioengineering and Biosciences, Lovely Professional University, Phagwara, Punjab, India
[2]Department of Botanical and Environmental Sciences, Guru Nanak Dev University, Amritsar Punjab, India
[3]Division Botany, Department of Bio-Sciences, Career Point University, Hamirpur, Himachal Pradesh, India
[4]College of Horticulture and Forestry, Dr. Y. S. Parmar University of Horticulture and Forestry, Nauni, Solan (HP), Neri Campus (Himachal Pradesh), India

Abstract

Coriandrum sativum Linn. belongs to family *Umbelliferae* and is commonly recognized as 'Coriander.' It is an annual herb cultivated across India, China, Netherlands, Bangladesh, Italy, Central and Eastern Europe. It is an aromatic, herbaceous, glabrous and annual herb with lot of medicinal properties. In ancient times it has been utilized for traditional medication as an antibacterial agent, anti inflammatory and analgesic. The oil derived from *C. sativum* is used as fragrance embedded with therapeutic properties. The various parts of *C. sativum* contain α-pinene, γ-terpinene, p-cymene, monoterpenes, limpnene, borneol, coriandrin, citronellol, geraniol, flavonoids, camphor, dihydrocoriandrin and essential oils. Different parts of this valuable plant viz, flower, seeds, stems and leaves possess anti-oxidant, anti-mutagenic, anti-convulsant, anti-diabetic, sedative hypnotic, hepatoprotective, diuretic, anti-microbial, anti-protozoal, anxiolytic, anti-ulcer and anthelmintic activity.

* Corresponding Author's Email: vndu.gndu@gmail.com.

In: *Coriandrum sativum*
Editors: Mamta Pujari, Bharat Kapoor, Neeta Pande et al.
ISBN: 978-1-68507-842-3

The main constituents of *C. sativum* are isocoumarins, volatile compounds and flavonoids. The seed oil comprises geranyl acetate and linalool. Being widespread throughout the globe along with significant biological activities, *C. sativum* normally being used in cooking can also be exploitedin medicine. Current chapter offers comprehensive information on pharmacological, botanical and chemical aspect of coriander and delineates the curative and traditional uses of coriander to assess its effectiveness as a medicinal agent.

Keywords: biological activities, botanical, coriander, herb, medicinal, pharmacological, therapeutic

1. Introduction

Coriandrum Sativum Linn belongs to family *Umbeliferae* and consists of a dried fruit known as Dhaniya (Evans 2002). This branched, slender and glabrous herb is cultivated across India, providing aromatic fragrance upon rubbing and is a vital annual herb originated from Mediterranean (Vaidya and Gogte 2000). This herbaceous plant grows upto 25-60 cm tall. The structure of plant is thin with upright stalk, spindle shaped roots, tiny white - violet tinted flowers and has alternate leaves (Figure 1 and Figure 2). Flowering of plant occurs during the month of June and July yielding round fruits containing two pericarps (Burdock and Carabin 2009). The fresh green leaves of *C. sativum* is called as cilantro and are used as flavoring agent in various cuisines (Wangensteen et al., 2004). Although the whole plant is edible but most common edible parts are dried seeds and leaves. In India it is commonly used in cooking as flavoring agent and is mostly found in Andhra Pradesh, Bihar, Himachal Pradesh, Karnataka, Madhya Pradesh, Rajasthan and Tamil Nadu (Handa and Maharaj 1996).

Herbal medicine is ancient form of therapeutics acknowledged by humankind. It was the backbone of primitive civilization and even in today's modern times it is being practiced extensively. Plants produce various phytochemical; primary metabolites and secondary metabolites and these are a vital source for numerous pharmaceuticals (Duke 2002; Gruenwalded 2004; I.H.P.R.N. 2002). On phytochemical screening, the coriander exhibits terpenoids, alkaloids, tannins, flavonoids, sterols, reducing sugars, phenolics, glycosides, essential oils and fatty acids. It also possesses huge nutritional biomolecules such as carbohydrates, minerals, vitamins, proteins, fibers, oils and trace elements. In Indian traditional medicine, coriander is utilized to treat respiratory, urinary and digestive disorders, because it contains diaphoretic, carminative, stimulant and diuretic properties. In Iranian medical science, coriander has been used to cure numerous health issues for example convulsion, dyspeptic problems, loss of appetite and insomnia (Benjumea et al., 2005; Maghrani et al., 2005; Mir 1992).

It has been reported that coriander exhibit many pharmacological effects including anti-diabetic activity (Matasyoh et al., 2009), sedative hypnotic activity (Emam and Heydari 2006), diuretic activity (Aissaoui et al., 2008), anti-mutagenic activity (Cortés-Eslava et al., 2004), anti-convulsant activity (Hosseizodeh 2015), anti-oxidant activity (Wangensteen et al., 2004), anthelmentic activity (Eguale et al., 2007), anti-fungal activity (Silva et al., 2011), cholesterol lowering activity (Dhanapakiam et al., 2007), anti-cancer activity (Chitra and Leelamma 2000), hepatoprotective activity (Pandey et al., 2011), post coital anti-fertility activity (Al-Said et al., 1987), anti-protozoal activity (Rondon et al., 2011), anxiolytic activity (Emamghoreishi et al., 2005), anti-ulcer activity (Al-Mofleha et al., 2006).

Figure 1. Figures showing leaves, flowers, fruits and seeds of coriander.

Table 1. Ayurvedic Description of *Coriandrum sativum* (Ayurvedic Pharmacopeia of India (2010); British Pharmacopoeia (2003); European Pharmacopoeia (2004)

Narrative of *Coriandrum sativum*	
Botanical name	*Coriandrum sativum* Linn.
Sanskrit name	Dhanika, Dhaniya, Vitunnaka, Kusutumbum
Synonyms	Dhana, Havija, Malli
Properties	
Rasa	Madhur, Tikta, Kashaya
Guna	Laghu, Snigdha
Virya	Ushna
Vipaka	Madhur
Doshaghnata	Tridoshahar, Snigdha
Karma (Actions)	Ushna = vatashaman; Kashaya, tikta, madhur = pittashaman Tikta, katu, ushna = kapashman Green coriander is pittashmak.
Curative uses	External use: General swelling and pain; Nasal drops; Lynphadenopathy; Headache; Burning sensation; Conjuctivitis; Haemostatic; Stomatitis; Internal use: CNS-Tonic for majjadhatu; syncope; memory loss; vertigo Digestive system: anti-dyseptic; liver stimulant; digestive; astringent; appeptizer; anthelmentic CVS: Bleeding disorders Respiratory system: Cough Dyspnoe Urinary system: diuretic, beneficial for prameha.

2. Botanical Depiction

Botanical depiction (Handa and Maharaj 1996; British Pharmacopoeia 2003; United states Pharmacopoeia 2004; European Pharmacopoeia 2004) (shown in Figure 1).

Table 2. The chemical composition of *Coriandrum sativum* (Diederichsen Axel 1996)

Component	Percentage (%)
Water	11.37
Fat	19.15
Starch	10.53
Protein	11.49
Fiber	28.43
Minerals	4.98
Pentosans	10.29
Sugar	1.92
Essential oil	0.84

2.1. Leaves

Numerously branched along with sub branches. Aerial old leaves are elongated while as new leaves are oval shaped.

2.2. Flowers

White coloured with slight violet shade.

2.3. Fruit

Round or oval shaped dicot seed, green in colour during early stage later upon maturation turns into beige colour.

2.4. Seed

Flowers and seeds produced during winter-end.

2.5. Macroscopic Representation

Fruit is globular whereas, mericarps are generally unified from their margins and are assembled into cremocarp which has a diameter of nearly 2-4 mm. The fruit is evenly brown or brownish yellow coloured, glabrous and in few cases covered by style and sepal residues. It has 10 primary ridges and 8 secondary ridges which are wavy and moderately inconspicuous. It is aromatic, spicy herb with characteristic taste (United states Pharmacopoeia 2004; Handa and Maharaj 1996).

2.6. Phytochemicals

The prevalent chemical compositions existing in coriander are delineated in Table 2. Fresh green coriander holds water upto 84% and 1.8% volatile oil is present in seeds. The main active components present in *Coriandrum sativum* L. are fatty oils and essential oils. In ripe and dried fruits, the essential oil content ranges between 0.03-2.6% whereas, fatty oil content ranges between 9.9-27.2%. The most vital component is linalool which is present in concentration between 60-70% and the other subsidiary components present are monoterpenes hydrocarbons such as α-pinene, γ-terpinene, borneol, geraniol acetate camphor, limpnene, geraniol, 0-cymene and citronellol, heterocyclic components such as pyridine, neochdilide, coriandrin, flavonoids, sterols, pyrazine, isocoumarins, thiazole, dihydrocoriandrin, pthlides, coriandrons A-E, digustilide phenolic acids, furan and tetrahudrofuran derivatives (Diederichsen Axel 1996; The Indian herbal pharmacopeia 2002).

3. Distribution

Coriandrum sativum L. is indigenously distributed in Eastern Mediterranean and is grown as a spice herb in Netherlands, China, Bangladesh, Egypt, India, Morocco, Central and Eastern Europe (Small 1997). However, in today's time it is being cultivated in Europe – Ireland, Sweden, Belgium, Spain, Denmark, Norway, United Kingdom, Germany, Austria, Hungary, Finland, Latvia, Poland, Belarus, Czechoslovakia, Netherlands, Estonia, Switzerland, Bulgaria, Ukraine, Yugoslavia, Greece, France, Lithuania, Romania, Albania, Moldova, Italy and Portugal; Northern Africa – Tunisia, Ethiopia, Morocco and Algeria; Asia – India, Iran, Georgia, Afghanistan, Pakistan, China, Iraq, Azerbaijan, Turkey, Palestine, Syria, Kazakhstan, Turkmenistan, Lebanon, Kyrgyzstan, Uzbekistan, Tajikistan, Jordan, Armenia and southern Russia (USDA, ARS, National Genetic Resource Program. GRIN, National Germplasm Resource Laboratory, Beltsville, Maryland, 2015; The University of Queensland 2011). Ukraine is top producer of coriander oils and governs the global price on demand basis. In India it is mainly grown in Andhra Pradesh, Himachal Pradesh, Bihar, Madhya Pradesh, Rajasthan, Tamil Nadu and Maharashtra (Ayurvedic Pharmacopeia of India 2010; European Pharmacopoeia 2004; British Pharmacopoeia 2003; Handa et al., 1996; United states Pharmacopoeia 2004).

4. Traditional Practices

Since 1550 BC the coriander came into use and was among the oldest spice in our planet. Therapeutically it was utilized as carminative, stimulant and aromatic purposes (Coskuner and Karababa 2007). Earlier, coriander was regarded as leading oil plant throughout the globe (Lawrence 1993). The primary products of coriander including fruits and leaves are utilized for traditional purposes as curative and culinary. Coriander is a common spice which is finely crushed and used as a main ingredient in many cuisines (Wangensteen et al., 2006). The seeds are also used to cure indigestion, convulsion, rheumatism, insomnia, worm infection, joint pain, loss of appetite and anxiety (Eikani et al., 2007; Masaada et al., 2007). Traditionally, in Morocco the coriander is utilized as diuretic plant (Aissaoui et al., 2008). In Iranian conventional medicine it is used to treat insomnia and anxiety (Emamghoreishi and Khasaki 2005). Also, it is conventionally used as a carminative, galactagogue, spasmolytic and digestive, whereas the seed extract is used in shampoos and lotions as an anti-microbial (Asolkar et al., 1992; Chopra et al., 1956; Ghani 1998). In addition, grounded seeds were applied to reduce swelling and pain and the paste made from green leaves was used externally to relief headache in earlier times. Powered coriander has been used externally to reduce pain and burning sensation in cases like inflammation lead by lymphadenopathy and erysipelas. Nasal drop made from green coriander works as haemostat and thereby prevent bleeding in cases like epistaxis. Extract and juice of coriander cures conjunctivitis. Internally coriander is utilized as a tonic. It has also been used as a treatment for memory loss and syncope. A gargle by fresh juice of coriander plant relieves sore throat and stomatitis (Craker and Simon 2002; Kirtikar and Basu 1999; Kokate et al., 2007).

5. Pharmacological Properties

5.1. Anti-Oxidant Activity

The extract of coriander procured by sequential extraction method lead to the identification of phenolic compounds which is accountable for its antioxidant action. The principal compounds viz, epoxide, neoxanthin, 6-epoxide, b-carotene, lutein-5, violaxanthin and b-cryptoxanthin were isolated from *C. sativum* via column chromatography and detected as per their spectral and chromatographic features. Extract from coriander oil, seeds and leaves with different polarity were examined for their anti-oxidant activity and might have a great potential as natural anti-oxidant that obstructs excess oxidation processes. Extracts derived from leaves and seeds displayed a concentration dependent DPPH scavenging activity (de Almeida Melo et al., 2005; Wangensteen et al., 2004).

5.2. Anti-Diabetic Activity

Following single dose of coriander extract at a concentration of 20 mg/kg in sub-chronic OHH-Meriones shawi rats, the glycemia got normalized and reduction in insulin levels, LDL cholesterol, triacyleglyceride, elevated insulin resistant and total cholesterol was observed without any marginal effect on body weight, creatinine and urea. Results also depict that upon regular consumption of coriander seeds the hyperglycemia gets lowered and the cardiovascular problems caused by hyperlipidemia or dyslipidemia can be prevented (Aissaoui et al., 2011). The anti-hyperglycemic activity of *Coriamdrum sativum* L. is related with inducing insulin secretion and increasing glucose uptake and metabolism by body muscle. Coriander included diet (62.5 g/kg) and drinking concoction prepared by boiling 2.5 g/l coriander for 15 minutes lowered the diabetes of streptozotocin-diabetic mice. Coriander juice 1 mg/ml revealed 1.6-fold enhancement in 2-deoxyglucose transport and 1.4 fold enhancement in glucose oxidation. Moreover, in 20 minutes of test, coriander extract at concentration of 0.25 to 10.00 mg/ml elicited 1.3-5.7-fold insulin secretion from clonal B-cell line. Coriander can emerge as a possible anti-diabetic dietary adjunct and a potential agent for diabetes therapy (Gray and Flatt 1999; Selvan 2003).

5.3. Anti-Microbial Activity

Aqueous decoction and extract of coriander were checked against 186 bacterial species which belongs to 10 diverse genera of gram-positive bacterial family and 2 isolates of *Candida albicans* obtained from urine samples (Sabahat and Perween 2007). The vital oil of coriander leaves extracted by hydro-distillation was examined. The main constituents obtained were 2E-decenal ~ 15.9%, decanal ~ 14.3%, 2E-decen-1-ol ~ 14.2% and n-decanol ~ 13.6%. Some other constituents were present in lesser amount including 2E-trideceen-1-al ~ 6.75%, 2E-dodecenal ~6.23%, dodecanal ~ 4.36%, undecanol ~3.37% and undecanal ~3.23%. The essential oil was tested for anti-microbial action against gram positive bacteria including *Staphylococcus aureus, Bacillus* spp. gram negative bacteria including *Escherichia coli, Salmonella typhi, Klebsiella pneumonia, Proteus mirabilis, Pseudomonas aeruginosae* and also against pathogenic fungus *Candida albicans* (Matasyoh et al., 2009).

5.4. Anthelmintic Activity

Effect of raw aqueous, hydro alcoholic extract of coriander seeds on egg and adult nematode parasite namely *Haemonchus contortus* for anthelmintic action were analyzed. Also, aqueous extract of coriander was examined for anthelmintic action *in vivo* conditions against sheep infected with parasite *Haemonchus contortus*. Both extracts of coriander inhibited egg hatching entirely at 0.5 mg/ml concentration. The aqueous extract was implemented at concentration of 0.12 mg/ml and hydro alcoholic extract at 0.18 mg/ml concentration. The hydro alcoholic extract proved more beneficial against adult parasite in *in vitro* conditions as compared to aqueous extract. Crude extract (hydro alcoholic and aqueous) of coriander seeds entirely obstructed nematode egg hatching at concentration of ~ 0.5 mg/ml. Although, hydro alcoholic decoction showed much higher *in vitro* action against parasites compared to aqueous extract. *In vivo* the anthelmintic action was checked by faecal egg count reduction and total worm count reduction in sheep which was artificially infected with *Haemonchus contortus*. Faecal egg count reduction was discerned on second day after treatment with 0.9 g/kg of crude aqueous decoction of coriander. On 7th and 14th day faecal egg count reduction was also observed at 0.45 g/kg dose of crude aqueous decoction and significant results were seen. Similarly, total worm count reduction was also discerned after treatment of 0.9 g/kg of crude aqueous decoction (Eguale et al., 2007).

5.5. Anti-Mutagenic Activity

The 4-nitro-o-phenylenediamine (NOP) is prominent mutagen whose mutagenic efficiency can be improved by plant metabolism. M-phenylenediamine (m-PDA) is transformed to mutagenic components identified by *Salmonella typhimurium* TA98 strain. 2-aminofluorene (2-AF) is another pro-mutagen which is plant activated and comprehensively studied. Plants activate 2-AF as well as m-PDA into powerful mutagens giving rise to DNA frame shift mutations. Coriander is a conventional plant used in Mexican diet, mostly consumed uncooked (Momin et al., 2012). In a study anti-mutagenic action of *Coriandrum sativum* L. extract was examined against mutagenic action of NOP and 2-AF using Ames reversion mutagenicity assay with *Salmonella typhimurium* TA98 acting as indicator organism. The co-incubation assay comprising plant cell or microbe was used as activating system for aromatic conversion and plant extract interaction. The crude aqueous extract of coriander remarkably reduced the mutagenicity of metabolized aromatic amines (AA) in the order as: 2-AF (92.43%) > m-PDA (87.14%) > NOP (83.21%) (Cortes-Eslava et al., 2004). The potential of *C. sativum* essential oil to stimulate nuclear DNA damage responsive genes was studied by exploring suitable Lac-Z fusion strains for the genes which are engaged in DNA metabolism and repair (RNR3 and RAD51) respectively. The coriander essential oil showed considerable gene induction, nearly same as that resulted by hydrogen peroxide, yet lesser than that triggered by methyl methanesulfonate. The coriander essential oil impairs mitochondrial structure, its function and can induce transcriptional expression of DNA damage responsive genes. It was manifested that microbial damage was directly linked to cellular cyto-toxicity by coriander essential oil which also concealed the nuclear genetic event occurrence (Bakkali et al., 2005).

5.6. Anti-Cancer Effect

The anti-cancer action of coriander (leaf, stem and roots) and its impact on cancer cell migration, DNA damage protection with main focus on roots was examined. Ethyl acetate concoction of coriander roots revealed maximum anti-proliferative action on MCF-7 cells (IC_{50} = 200 ± 2.6 μg/ml) with maximum phenolic content and FRAP and DPPH scavenging activity compared to other coriander extracts. The ethyl acetate concoction of coriander root prevented DNA damage and MCF-7 cell migration caused by hydrogen peroxide, proposing its possibility in cancer deterrence and preventing metastasis. The extract possesses anti-cancer activity in MCF-7 cells by influencing antioxidant enzymes probably resulting in hydrogen peroxide accumulation, cell cycle arrest (G2/M phase) and apoptotic cell death via death receptors and mitochondrial apoptotic pathways (Tang et al., 2013). The anti-tumor and immune-regulating action of aqueous and methanolic extract of coriander (seed and leaf) was studied in *in vitro* conditions. The aqueous concoction of coriander leaf resulted in ($P < 0.05$), 24% to 39% L5178Y-R lymphoma cell toxicity at concentration of 31.2 μg/ml (MIC), while the methanolic concoction of coriander leaf and seed resulted in 31% and 40% cyto-toxicity at concentration 62.5, 7.8 μg/ml (MICs) respectively.

Additionally, aqueous concoction from coriander leaf caused ($P < 0.01$) 14% – 45% splenic cells lympho-proliferation at 125 μg/ml concentration. Further, coriander aqueous concoction was significant ($P < 0.01$) andreduced ~ 100% nitric oxide generation by LPS stimulated macrophages (Gomez-Flores et al., 2010). Three diverse cell lines were explored to find out cytotoxicity which includes BMK – kidney, WRL-68 – liver and KHOS-2405 – bone. Different concentration of coriander treatments was given to cells that is, 0.125%, 0.25%, 0.5%, 1%, 1.5%, 2% and 2.5%) for 24 h. All the three cell lines showed reduced proliferation and number of cells corresponding to the treatment concentrations. Coriander arrested cell cycle of BMK cells at (G2) and (M) phase, WRL-68 cells at (S) phase and KHOS cells at (G1) phase (Rodriguez et al., 2006).

5.7. Anti-Inflammatory and Analgesic Effect

The anti-inflammatory and anti-granuloma action of coriander extract (hydro-alcoholic) was investigated in experimental models. The anti-inflammatory action of coriander hydro-alcoholic extract was checked using carrageenan induced paw edema model and anti-granuloma action of coriander hydro-alcoholic extract was checked using subcutaneous cotton pellet implantation stimulated granuloma formation and regulation of peritoneal macrophages with complete Freund's adjuvant. Serum tumor necrosis factor-α (TNF-α), IL-1 β, IL-6 levels and peritoneal macrophage expression of TNF-R1 were estimated as markers of global inflammation. Coriander hydro-alcoholic extract at concentration of 32 mg/kg caused ($P < 0.05$) decrease in paw edema following carrageenan administration. The treatment of coriander hydro-alcoholic extract decreased dry granuloma weight in experimental animals. Serum IL-1 β, IL-6 levels were significantly ($P < 0.05$) reduced in coriander treated (32 mg/kg) animals comparative to controls. However, an increase was observed in serum TNF-α levels in treated models comparative to the controls, yet TNF-R1 expression was declined in peritoneal macrophages (Nair et al., 2013). The anti-inflammatory and analgesic impact of coriander seeds was studied in animal models. Anti-inflammatory action was assessed using carrageenan test whereas for checking analgesic action formalin and writhing test was used. In writhing test, the coriander essential oil has remarkable impact ($P < 0.01$). Total concoction, polyphenolic extract and essential oil has considerable impact in two phases of formalin test (Haj et al., 2003).

Conclusion

Throughout the history, mankind has discovered that certain herbs and plants not only enrich the food flavor but also revitalize health. The current chapter discusses about advances in using *Coriandrum sativum* L. as phyto-therapy and also to demonstrate its potential as therapeutic agent. From current information it is apparent that *Coriandrum sativum* L. possesses pharmacological effects such as antioxidant, anti-diabetic, anti-microbial, anti-cancer, anti-mutagenic and anti-inflammatory. The therapeutic properties of *Coriandrum sativum* L. can be attributed to its remarkable phytonutrients content. The present literature manifests *Coriandrum sativum* L. as medicinal herb. The literature also shows the possibility of coriander fruit, essential oil or extract useful for development of new pharmaceutical or herbal medicines to treat many diseases. For future studies comprehensive, pharmacological and chemical experiments including human metabolism should be considered and given full focus in order to benefit humankind.

References

Aissaoui, A., Jaouad El-Hilaly, Zafar H. Israili, and Badiâa Lyoussi. 2008. "Acute diuretic effect of continuous intravenous infusion of an aqueous extract of *Coriandrum sativum* L. in anesthetized rats." *Journal of ethnopharmacology* 115: 89-95.

Aissaoui, A., Soumia Zizi, Zafar H. Israili, and Badiâa Lyoussi. 2011. "Hypoglycemic and hypolipidemic effects of *Coriandrum sativum* L. in Meriones shawi rats." *Journal of Ethnopharmacology* 137: 652-661.

Al-Mofleh, I. A., A. A. Alhaider, J. S. Mossa, M. O. Al-Sohaibani, S. Rafatullah, and S. Qureshi. 2006. "Protection of gastric mucosal damage by *Coriandrum sativum* L. pretreatment in Wistar albino rats." *Environmental Toxicology and Pharmacology* 22: 64-69.

Al-Said, M. S., K. I. Al-Khamis, Mohammad W. Islam, N. S. Parmar, M. Tariq, and A. M. Ageel. 1987. "Post-coital antifertility activity of the seeds of *Coriandrum sativum* in rats." *Journal of ethnopharmacology* 21: 165-173.

Asolkar, L. V., Kakkar, K. K., & Chakre, O. J. 1992. Glossary of Indian medicinal plants with active principles (part 1). *New Delhi: National Institute of Science Communication And Information Resources (CSIR)*, 230-231.

Bakkali, F., S. Averbeck, D. Averbeck, Abaudoux Zhiri, and M. Idaomar. 2005. "Cytotoxicity and gene induction by some essential oils in the yeast *Saccharomyces cerevisiae*." *Mutation Research/Genetic Toxicology and Environmental Mutagenesis* 585: 1-13.

Benjumea, D., S. Abdala, F. Hernandez-Luis, P. Pérez-Paz, and D. Martin-Herrera. 2005. "Diuretic activity of Artemisia thuscula, an endemic canary species." *Journal of Ethnopharmacology* 100: 205-209.

British pharmacopoeia, Introduction General Notices Monographs, medicinal and Pharmaceutical, *British pharmacopeia commission*, London, Volume-1 (A-I), 2003: 542-43.

Burdock, G. A., and Ioana G. Carabin. 2009. "Safety assessment of coriander (*Coriandrum sativum* L.) essential oil as a food ingredient." *Food and Chemical Toxicology* 47: 22-34.

Chithra, V., and S. Leelamma. 2000. "*Coriandrum sativum* - effect on lipid metabolism in 1, 2-dimethyl hydrazine induced colon cancer." *Journal of Ethnopharmacology* 71: 457-463.

Chopra, R. N., Nayar, S. L., Chopra, I. C., Asolkar, L., & Kakkar, K. 1956. Glossary of Indian medicinal plants, New Delhi. *Council of Scientific & Industrial Research, s.*

Cortés-Eslava, J., Sandra Gómez-Arroyo, Rafael Villalobos-Pietrini, and Jesús Javier Espinosa-Aguirre. 2004. "Antimutagenicity of coriander (*Coriandrum sativum*) juice on the mutagenesis produced by plant metabolites of aromatic amines." *Toxicology letters* 153: 283-292.

Coşkuner, Y., and Erşan Karababa. 2007. "Physical properties of coriander seeds (*Coriandrum sativum* L.)." *Journal of Food Engineering* 80: 408-416.

Craker, L. E. and Simon J. E. *Herb spices and medicinal plant*, vol. 3, CBS Publishers and Distributors, New Delhi 2002.

de Almeida Melo, E., Jorge Mancini Filho, and Nonete Barbosa Guerra. 2005. "Characterization of antioxidant compounds in aqueous coriander extract (*Coriandrum sativum* L.)." *LWT-Food Science and Technology* 38: 15-19.

Dhanapakiam, P., J. Mini Joseph, V. K. Ramaswamy, M. Moorthi, and A. Senthil Kumar. 2007. "The cholesterol lowering property of coriander seeds (*Coriandrum sativum*): mechanism of action." *Journal of Environmental Biology* 29: 53.

Diederichsen, A. 1996. *Coriander: Coriandrum sativum L* (Vol. 3). Bioversity International.

Duke, J. A. 2002. *Handbook of medicinal herbs*. CRC press.

Edition, *I. H. P. R. N.* 2002. Published by Indian Drug Manufacturer's Association.

Eguale, T., G. Tilahun, A. Debella, A. Feleke, and E. Makonnen. 2007. "*In vitro* and *in vivo* anthelmintic activity of crude extracts of *Coriandrum sativum* against Haemonchus contortus." *Journal of Ethnopharmacology* 110: 428-433.

Eikani, M. H., Fereshteh Golmohammad, and Soosan Rowshanzamir. 2007. "Subcritical water extraction of essential oils from coriander seeds (*Coriandrum sativum* L.)." *Journal of Food Engineering* 80: 735-740.

Emam, G. M., and Hamedani G. Heydari. 2006. "*Sedative-hypnotic activity of extracts and essential oil of coriander seeds*." 22-27.

Emamghoreishi, M., Mohammad Khasaki, and Maryam Fath Aazam. 2005. "*Coriandrum sativum*: evaluation of its anxiolytic effect in the elevated plus-maze." *Journal of ethnopharmacology* 96: 365-370.

Evans, W. C. 2002. *Trease and Evans: Pharmacognocy*. Fifteenth International Edition.

Ghani, A. 1998. *Medicinal plants of Bangladesh: chemical constituents and uses*. Asiatic society of Bangladesh.

Gomez-Flores, R., Humberto Hernández-Martínez, Patricia Tamez-Guerra, Reyes Tamez-Guerra, Ramiro Quintanilla-Licea, Enriqueta Monreal-Cuevas, and Cristina Rodríguez-Padilla. 2010. "Antitumor and immunomodulating potential of *Coriandrum sativum*." *Journal of Natural Products* 3: 54-63.

Gray, A. M., and Peter R. Flatt. 1999. "Insulin-releasing and insulin-like activity of the traditional anti-diabetic plant *Coriandrum sativum* (coriander)." *British Journal of Nutrition* 81: 203-209.

Gruenwalded, J. 2004. PDR-HM Physicians' desk reference for herbal medicine. *Medical Economics. NJ*, *8*, 378-84.

Haj, H. V., Ghannadi, A., & Sharif, B. 2003. *Anti-inflammatory and analgesic effects of Coriandrum sativum L. in animal models*.

Handa, S. S., and Maharaj Krishen Kaul. 1996. "*Supplement to cultivation and utilization of medicinal plants*."

Hosseizodeh, H. "Anticonvulsant Effect of *Coriandum sativum* L. *Seed Extract in Mice* 2011." (2015).

Kirtikar, K. R. and Basu B. D. *Indian medical plants, second edition,* vol. 2, International Book Distributers, Dehradun, India 1999.

Kokate, C. K., Purohit A. P. and Gokhale S. B. *Pharmacognosy,* 39th edition, Nirali Prakashan, Pune 2007.

Lawrence, B. M. 1993. "A planning scheme to evaluate new aromatic plants for the flavor and fragrance industries." *New crops* 1: 620-7.

Maghrani, M., N. A. Zeggwagh, M. Haloui, and M. Eddouks. 2005. "Acute diuretic effect of aqueous extract of Retama raetam in normal rats." *Journal of Ethnopharmacology* 99: 31-35.

Matasyoh, J. C., Z. C. Maiyo, R. M. Ngure, and R. Chepkorir. 2009. "Chemical composition and antimicrobial activity of the essential oil of *Coriandrum sativum*." *Food Chemistry* 113: 526-529.

Mir, H. 1992. *Coriandrum sativum. Application of plants in prevention and treatment of illnesses. Persian*, *1*, 257-52.

Momin, A. H., Sawapnil S. Acharya, and Amit V. Gajjar. 2012. "*Coriandrum sativum*-review of advances in phytopharmacology." *International Journal of Pharmaceutical Sciences and Research* 3: 1233.

Monograph of the fifth edition of European Pharmacopoeia 2004; *Stationary office on behalf of the medicines and healthcare products Regulatory agency (MHRA)*; London: The stationary office, 2008: 617.

Msaada, K., Karim Hosni, Mouna Ben Taarit, Thouraya Chahed, Mohamed Elyes Kchouk, and Brahim Marzouk. 2007. "Changes on essential oil composition of coriander (*Coriandrum sativum* L.) fruits during three stages of maturity." *Food Chemistry* 102: 1131-1134.

Nair, V., Surender Singh, and Yogendra Kumar Gupta. 2013. "Anti-granuloma activity of *Coriandrum sativum* in experimental models." *Journal of ayurveda and integrative medicine* 4: 13.

Pandey, A., P. Bigoniya, V. Raj, and K. K. Patel. 2011. "Pharmacological screening of *Coriandrum sativum* Linn. for hepatoprotective activity." *Journal of Pharmacy and Bioallied sciences* 3: 435.

Rodríguez, L., Mariana Ramirez, Ana Badillo, Angel León-Buitimea, and Jorge Reyes-Esparza. 2006. "*Toxicological evaluation of Coriandrum sativum (Cilantro) using in vivo and in vitro models.*" A645-A645.

Rondon, F. C. M., Claudia M. L. Bevilaqua, Marina P. Accioly, Selene M. Morais, Heitor F. Andrade-Junior, Lyeghyna K. A. Machado, Roselaine P. A. Cardoso, Camila A. Almeida, Eudson M. Queiroz-Junior, and Ana Caroline M. Rodrigues. 2011. "*In vitro* effect of Aloe vera, *Coriandrum sativum* and *Ricinus communis* fractions on *Leishmania infantum* and on murine monocytic cells." *Veterinary parasitology* 178: 235-240.

Saeed, S., and Perween Tariq. 2007. "Antimicrobial activities of *Emblica officinalis* and *Coriandrum sativum* against gram positive bacteria and *Candida albicans*." *Pak. J. Bot* 39: 913-917.

Selvan, M. T. 2003. Role of spices in medicine. *Indian J. Arecanut Spice. Medicinal Plant*, *5*, 129-135.

Silva, F., Susana Ferreira, Andreia Duarte, Dina I. Mendonca, and Fernanda C. Domingues. 2011. "Antifungal activity of *Coriandrum sativum* essential oil, its mode of action against Candida species and potential synergism with amphotericin B." *Phytomedicine* 19: 42-47.

Small, E. *Culinary herbs*. NRC Research Press, Ottawa 1997:219-225.

Tang, E. L. H., Jayakumar Rajarajeswaran, Shin Yee Fung, and M. S. Kanthimathi. 2013. "Antioxidant activity of *Coriandrum sativum* and protection against DNA damage and cancer cell migration." *BMC complementary and alternative medicine* 13: 1-13.

The Ayurvedic Pharmacopeia of India. Government of India, Ministry of Health and family warfare department of Indian system of medicine and Homeopathy, first edition, Part-1, volume-1, the controller of publications, civil lines, Delhi, India, 2010: 30-31.

The Indian herbal pharmacopeia. Revised edition, published by Indian drug manufacturer association, Mumbai, India, 2002: 144-52.

The University of Queensland. *Special edition of environmental weeds of Australia for biosecurity Queensland*, http://keyserver.lucidcentral.org/weeds/data/080c0106-040c-4508-8300-0b0a06060e01/media/Html/Conium_maculatum.htm (2011).

USDA, ARS, National Genetic Resources Program. *Germplasm Resources Information Network-(GRIN)*. National Germplasm Resources Laboratory, Beltsville, Maryland. URL: http://www.ars-grin.gov.4/cgi-bin/npgs/html/taxon.pl?11523 (22 July 2015).

USP 27-The United States pharmacopoeia, & NF 22- The National formulary; Asian edition; by authority of the United States pharmacopial convention, Inc., Washington, 2004: 2853

Vaidya, V. M., and M. Gogte. 2000. "Ayurvedic Pharmacology & Therapeutic uses of medicinal plants." *Swami Prakashananda ayurvedic research center, Mumbai.* 656-657.

Wangensteen, H., Anne Berit Samuelsen, and Karl Egil Malterud. 2004. "Antioxidant activity in extracts from coriander." *Food chemistry* 88: 293-297.

Chapter 7

Disease Management Strategies and Importance of *Coriandrum sativum* L.

Monika Thakur[1], Pungbili Islary[2], Bharat Kapoor[3] and Mamta Pujari[4,*]
[1]Division of Botany, Department of Bio-Sciences, Career Point University, Hamirpur (Himachal Pradesh), India
[2]Department of Botany, Bodoland University, Kokrajhar (Assam), India
[3]Department of Hotel Management and Tourism, Guru Nanak Dev University, Amritsar (Punjab), India
[4]Department of Botany, School of Bioengineering and Biosciences, Lovely Professional University, Phagwara (Punjab), India

Abstract

Coriandrum sativum L. is a prominent species belonging to the *Apiaceae* family and therefore is popularly used as a condiment, and also in the pharmaceutical and food industries. In addition to being used as a flavoring in preparing food, the plant's leaves and seeds are frequently employed in ethnomedicine. It must be primarily grown for the volatile oils, fatty acids, terpenoids, flavonoids, as well as polyphenols found in the seeds (fruits). In various ailments, the fruits have therefore internally carminative, spasmolytic, as well as galactagogic properties. Coriander is being used in a variety of ways, including fresh and dried herbs. As diverse chemical constituents in various parts of the plant, *C. sativum* L. essential oil, as well as extracts, have significant antimicrobial, antifungal, or anti-oxidative effects, and hence play a vital role in preserving the storage of assets by inhibiting contamination. Because this edible plant is non-toxic to living beings, *C. sativum* L. essential oil is

* Corresponding Author's Email: mamta.21901@lpu.co.in.

In: *Coriandrum sativum*
Editors: Mamta Pujari, Bharat Kapoor, Neeta Pande et al.
ISBN: 978-1-68507-842-3

utilized in a variety of applications, including flavoring agents and preservation, medicinal applications (therapeutic activity), as well as aromas (fragrances and lotions). The current chapter includes all aspects of *C. sativum* L., including its origin, taxonomic status, distribution, botanical description, cultivation, as well as important diseases that influence it, with an emphasis on its economic potential.

Keywords: *Coriandrum sativum*, *Apiaceae*, taxonomy, essential oil, cultivation

1. Introduction

C. sativum L., is an herbaceous plant in the *Apiaceae* (*Umbelliferae*) family which has been widely harvested in temperate regions for centuries, including the Middle East, Latin America, Africa, and Asia. Coriander seeds, as well as aerial components, have frequently been utilized as food ingredients. Fresh coriander leaves and seeds, as well as dried coriander seeds, are used to flavour food (Hwang et al. 2014). Essential oils as well as monoterpenoids, including linalool, are the major constituent elements in coriander seeds. *C. sativum* L. is a coriaceous fragrant, shrubby annual herb with a strong presence as a flavoring agent. It is a source of aromatic compounds and essential oils with bioactive molecules that have antibacterial, antifungal, and antioxidant activities, producing *C. sativum* L. effective in food processing (as a flavor enhancer as well as adjuvant) and restoration or in helping to prevent food contamination. *C. sativum* L. consists of two types of herbal raw products: fruits and leaves, with essential oils as the key physiologically main substance. Coriander seeds are being used as a fragrant flavor in cuisines while also serving as a dietary supplement, significantly improving the digestion process. The quantity of *C. sativum* L. essential oils, as well as their chemical composition, alter throughout morphogenesis (Bhuiyan et al. 2009), affecting the plant's aroma, and therefore the coriander fruit flavor differs significantly from the herb's aroma (Neffati and Marzouk 2008).

The odor of immature fruits and leaves is referred to as a "stink insect scent" and is caused by the presence of trans-tradesmen in the oil. In the Sanskrit language, coriander is known as "kusthumbari" or "dhanayaka"; in Hindi, it is known as Dhania, and in Bengali, it is known as Dhane. It is an eastern Mediterranean native that has been distributed to India, China, and the entire world (Coşkuner and Erşan 2007). Coriander is utilized as a condiment, remedy, and raw ingredient in the beverages industry, or in pharmaceutical products. The mature fruits have a fresh, delightful aroma and have been widely utilized in-ground and volatile extract from throughout the globe. The essential oil content of coriander fruits has been studied in many different locations of the world and has been found to vary. The coriander plant produces two main components that are used to enhance the flavor: a fresh green herb as well as a condiment (mature dried seed capsule or fruit). These raw materials have very unique aromas and tastes. In addition to its main uses as a seasoning as well as therapeutic herb, the herb is utilized as a flavor enhancers agent in food, fragrances, perfumes, as well as cleansers, which gives it economic benefit (Burdock and Ioana 2009; Ravi et al. 2007). It is frequently utilized in Indian cuisine however as a remedy in Indian medical systems. From the outset, it has been greatly valued across native remedies, notably numerous medical traditions.

2. Origin and Distribution

Coriander is known to have existed in the Mediterranean region. Only one species, *C. sativum* L., is frequently harvested, primarily in the tropics. India is the leading producer in coriander yield and processing, with a total area of 5.25 hectares under production and average productivity of 3.10 tonnes. Andhra Pradesh, Rajasthan, Madhya Pradesh, Karnataka, Tamil Nadu, and Uttar Pradesh seem to be the primary coriander-growing states in India. Coriander is grown in Morocco, Romania, France, Spain, Italy, the Netherlands, Myanmar, Pakistan, Turkey, Mexico Argentina, and, to a lesser extent, the United Kingdom and the United States (Khan et al. 20). There may be two different morphological forms: one is erect and tall, with a greater primary shoot as well as shorter branches, while the other is bushy, with a shorter main shoot as well as longer, extending branches. Depending on the species, the plants can grow to be 30 to 100 cm tall. Based on the variety as well as agricultural circumstances, the crop flowers 45–60 days after planting as well as matures in 65 days. Each branch, and also the main stem, culminates in a component umbel (evaluate growth) with 3–10 umbels and 10–50 pentamerous flowers in each umbel. Small, protandrous blooms are hard to analyze for pollination regulation. *C. sativum* L., like other umbelliferous herbs, is a cross-pollinated cultivar (Nadeem et al. 2013).

3. Taxonomy and Botanical Description

Coriander is a member of the *Apiaceae* family and the genus *Coriandrum*. Only two species of the herb are recognized: *C. sativum* L. as well as its wild related *C. tordylium*.

Vernacular Names:

Eng.: coriander, collender, Chinese parsley Hindi: dhania, dhanya Sanskrit: dhanayaka, kusthumbari Chinese: yuan sui, hu sui Fench: coriandre, persilarabe German: Koriander, Wanzendill, Schwindelkorn Italian: Coriandolo Spanish: coriandro, cilantro, cilandrio, culantro Turkish: Kisnis

Taxonomic Categorisation:

Kingdom: Plantae Division: Tracheophyta Class: Magnoliopsida Order: Apiales Family: Apiaceae Genus: *Coriandrum* Species: *sativum*

C. sativum L. is an annual herbaceous plant. The glabrous bush may grow to a height of 0.20 to 1.40 m when blooming. The plant seems to have a tap root as well as epigeal growth. The stem is tall or sympodial, monochasial-branched, as well as frequently contains several side branches at the basal node. Each branch seems to have an inflorescence at the end. The mostly ribbed stem is greenish as well as changes color in violet throughout shade during the flowering phase. The adult plant's stem is porous, and its basal regions may grow up to 2 cm in diameter. The leaves alternate, with the main ones frequently forming a rosette. The bush has a variety of leaves (Yeung et al. 2011). The blade form of the basal foliage is normally split having three lobes and tripinnatifid, while the leaves of the nodes below are pinnatifid to a larger extent. The more pinnate the leaves are, the greater these are implanted.

As a result, the upper leaves have narrow lanceolate or even filiform-shaped blades that are severely embossed. Lower leaves are stalked, whereas upper leaves' petioles have been limited to a tiny, virtually amplexicaul leaf.

The leaves are light green colored, with a glossy waxy surface. The leaves turn crimson or violet throughout the flowering phase. Beginning with the basal leaves, gradually fade before the fruits mature. A compound umbel is the type of inflorescence (Khan et al. 2014). There are one or two linear bracts in rarely. The umbel has two to eight main rays of varying lengths which have been arranged so that the umbellets are all at the same level. The umbellets are carried by two, three, and sometimes more bracteoles, each with five to twenty secondary rays. The major umbel is the first to mature. The periphery umbellets as well as umbellet flowers would be the first to mature across every umbel and umbellet. These blooms have a protandrous quality. The umbellets' core blossoms are staminiferous as well as sterile. The five calyx spikes enclosing the stylopodium have been evident in the ripe fruit of *C. sativum* L., which also has an inferior ovary (Mandal and Mandal 2015). The five calyx spikes, like the petals of flowers inside the periphery, possess varied lengths. There are five petals on each bloom. Because the petals on the outer of the umbellets are extended, the blooms on the periphery of each umbellet seem uneven. The blooms in the center are spherical and have little inflexed petals. The petals are pale pink or even sometimes whitish.

4. Chemical Composition

The chemical composition of coriander seeds is shown in Table 1. Essential oils, as well as fatty oil, are the most major components of *Coriandrum sativum* L. The volatile oil's composition of mature as well as dried coriander seeds ranges from 0.03 and 2.6 percent, whereas the fatty oil yield ranges from 9.9 and 27.7%. 1% of total S-(+)-linalool is the most essential aspect (60-70 percent) Monoterpene hydrocarbons might be other significant main compounds found in essential oils viz. α-pinene, limonene,γ terpinene,p-cymene, borneol, citronellol, camphor, geraniol and geraniol acetate, heterocyclic components like pyrazine, pyridine, thiazole, furan and tetrahydrofuran derivatives, isocoumarins, coriandrin, dihydrocoriandrin, coriandrons A-E, flavonoids, pthlides, neochidilide, digustilide phenolic acids and sterols (Diederichsen 1996; Pathak et al. 2011).

Table 1. Chemical composition of *C. Sativum* L

Component Content	Percentage (%)
Water	11.37
Crude protein	11.49
Fat	19.15
Crude fibre	28.43
Starch	10.53
Pentosans	10.29
Sugar	1.92
Mineral constituents	4.98
Essential oil	0.84

5. Cultivation

5.1. Climate

Despite coriander is a tropical plant, it may be cultivated in a variety of climates. It grows best in a cold atmosphere throughout the initial stages of development as well as matures in a warmer, drought environment. The optimum time to plant is between October and February. It can be planted in small beds throughout the year in urban areas for leaf yield. It is grown as an irrigated harvest throughout the months of June-July as well as September-October when the north-eastern rainy season begins, then cultivated when fully mature in January-February (Pathak et al. 2011).

5.2. Soils and Preparation of Land

Coriander might well be grown as an irrigated harvest on any soil type as long as enough organic fertilizer is provided. During rain-fed environments, black cotton soils having strong moist retentively are the greatest. With the advent of the monsoon, the field is ploughed three to four times for precipitation crops. After ploughing the area, beds, as well as pathways, are constructed for agricultural areas (Krishnan et al. 2001).

5.3. Manures and Fertilizes

In rain-fed regions, well-rotten farmyard manure (10-15 tonnes/ha) as well as fertilizers at rates of 20 kg nitrogen, 30 kg phosphorus, and 20 kg potassium per hector are used during crop cultivation before planting to ensure a productive crop. N dosage is increased to 60 kg/ha in irrigated areas, with half of the positive impact as a basal dose as well as the remaining portion given 30-45 days after planting (Czygan et al. 2001).

5.4. Sowing

Before planting, rub the seeds together to avoid splitting the fruits. Fruits and seeds grow a few days before the whole harvests. Soaking the seeds in water for 12-14 hours and then drying them in shade for 12 hours improves the seed germination. Seeds are sprayed with fungicides before planting. As a preventive strategy towards stem-gall infection, use Thiram at a dose of 2.5 g/kg of seeds. The seeds are planted in lines with a 25 cm row spacing as well as a 15 cm plant spacing. Seeds are often dispersed and then blended with soil with conveyors. Germination takes about 10-15 days, depending on the temperature (Czygan et al. 2001; Pathak et al. 2011).

5.5. Irrigation

Irrigation is based primarily on the soil, meteorological circumstances, as well as seasons. Very little irrigation is required for crops grown in black cotton soils, but 3-4 watering are needed for crops in light soils: the first at the 2-leave phase (20-30 days after planting); the second at the branching or blooming phase (60-70 days); and the third at the seed-filling phase (60-70 days) (80-110 days). A suitable level of soil humidity ought to be available to the plant at the period of blooming onset (Czygan et al. 2001; Pathak et al. 2011).

5.6. Intercultural

Weeding and hoeing are required for an effective crop. In most cases, two hoeing's are adequate for a healthy crop. The first is done when the plants are well above ground, as well as the second is done before the rows are rotted away; however, whether there is early precipitation during the cropping system, extra hoeing and weeding are done to eradicate weeds and improve soil porosity. For irrigation purposes, the first hoeing, as well as weeding, is done about 30 days of planting, although relying on weed development, one or two more weeding's are completed. (Czygan and colleagues, 2001).

6. Major Diseases and Preventive Measures

The crop suffers due to several biotic and abiotic stresses which are detrimental to plant health and seed quality. Several disease-causing pathogens are seed-borne. They are associated with seeds externally, internally, as a contaminant and associated with inert matter. Their seed to plant transmission may be local or systemic or both. It is essential to know the actual location of the pathogen with seeds for its management. Plant-to-seed infection and the timing of infection must be planned in order to prepare actions to ensure pathogen-free seed as yield. *C. sativum* L. is a significant herb crop farmed in a variety of nations, with India accounting for the largest share. Various pathogens infect the crop, causing diseases like vascular wilt, stem rot, root rot, charcoal rot, seedling blight, stem gall, leaf blight, anthracnose, powdery mildew, seed rot, grain mold, bacterial blight, soft rot, seedling root rot, reniform and root-knot nematodes, phyllody, and virus diseases (Khare et al. 2017) (Table 2). In several nations, new diseases had emerged. The greatest way to manage infections seems to be to grow tolerant varieties.

6.1. Studies of New Diseases in Coriander

Dennis (2003) described a novel infection caused by *Microdochium* sp. that was discovered in Australia. On the stems, the plants produced light grey, uneven deep infections up to one centimeter long. Branches and umbels were distorted, and the infection damaged the entire stem. The infection started in clusters, but eventually spread to several other crops, leading to a significant reduction in yield. *Plectoshaerella pauciseptata* causes soft rot signs on the stem as well as the petiole of coriander, which turns brown to black, according to Usmani et al. 2012. The vascular area became discoloured as well. *Plectoshaerella* rot has been the description referring to the infection. Rodeva et al. (2010) presented a research analysis on wilting as well as root rot induced by *Macrophomina phaseolina*. For the first instance, Garibaldi et al. (2010) documented coriander collar or root rot caused by *Pythium ultimum* in Italy. Cerkauskas discovered *A.* bacterial leaf spot on coriander in Ontario, Canada (2009).

Table 2. Diseases of coriander and their pathogens

Nature of pathogen	Pathogen	Disease
Fungi	*Fusarium oxysporum*	Vascular wilt
	Sclerotinia sclerotiorum	Stem rot
	Rhizoctonia solani	Root rot
	Rhizoctonia bataticola	Charcoal rot
	Alternaria poonensis, Pythium ultimum	Seedling blight
	Protomyces macrosporus	Stem gall
	Alternaria alternate	Leaf blight
	Cercospora corianderi, Phoma multirostrata	Leaf spot
	Colletotrichum capsici, C. gloeosporioides, Gloeosporium achaeniicola.	Anthracnose
	Erysiphe polygoni	Powdery mildew
	Fusarium miniliforme, F. semitectum, F. equiseti, Aspergillus spp.	Seed rot
	Helminthosporium, Fusarium, Curvularia, Alternaria	Grain mould
Bacterial	*Xanthomonas campestris, X. translucens*	Bacterial blight
	Erwinia carotovora	Bacterial sof rot
	Pseudomonas sp.	Seedling rot
Virus	-	Carrot red leaf virus
Nematodes	*Meloidogyne incognita*	Root knot

6.2. Management Resistance

The greatest strategy to manage infections in a crop would be to have types that are tolerant to disease-causing organisms. Methods have been developed, with some effectiveness, to produce such types available with tolerance to a specific disease and numerous infections (Malhotra and Vashishtha 2008). Accessions PH-7, Pant Haritma, CO-17, and CO-2 were shown to be particularly sensitive to stem gall by Singh et al. (2003). UD-03, UD-41, UD-46, and UD-48 are susceptible to stem gall, while G-C-88-8, G5365-91, Pant-1, and UD-20 are somewhat tolerant, according to Tripathi et al. (1998).

Powdery mildew resistance was observed by Keshwal and Khatri (1998) in varieties CS-362, CH-15, and CS-52. Varieties RCr CO-1 and CO-2 have been documented by Das (1998). *M. incognita* is sensitive to UD-21 as well as moderately resistant to RCr-41, CO-3, and UD 20.

6.3. Cultural Control

C. sativum L. must be cultivated in the winter in a frost-free, relatively cool, and drought place to reduce disease pressure. The soil ought to be medium to heavy loam with a pH of 6-7. Farming techniques are essential to stop soil-borne diseases. These diseases are held in place by soil solarisation. It is critical to sow early in the case of powdery mildew to avoid infection. Wilt could be controlled by using an oilcake or maintaining a pH of 8.2 in the soil (Khare et al. 2017). Potassium as well as nitrogen fertilizers, according to Tripathi et al. (2002), diminished stem gall occurrence. When determining the fertilizer concentration, caution is required. The defective parts of the plant must be eliminated, although if necessary, the entire plant should be uprooted and removed. Control using chemicals diseased seeds and inert debris must be eliminated from seed batches before they can be utilized as seed. Fungicides such as Mancozeb, captan, thiram, captafol, and Bavistin + thiram are required for seed treatment. Karathane should be used to manage powdery mildew. When the heat might not be too severe during the day, wettable sulfur can be applied. Mancozeb can be used to analyze different foliar infections. Dry warm air exposure at 53 °C for 30 minutes yields the best control of seed-borne bacterial infections (Champawat and Singh 2008). After 45 days of seeding, Singh (2009) found that applying *Trichoderma harzianum* and *Pseudomonas fluorescence* had been beneficial in reducing wilt.

7. Economic Importance and Uses of *C. sativum* L.

7.1. Traditional Uses

C. sativum L. has been utilized in many forms all over the globe. All components of the herb, such as the leaves, seeds, and essential oil, have long been employed in traditional medicines systems on the planet. The leaves are being used to relieve gastrointestinal irritations however as an antispasmodic, dyspeptic, and stimulant. Coughs, chest aches, bladder problems, and while an aphrodisiac being treated using leaves mixtures that are consumed as well as applied directly (Sahib et al. 2013). *C. sativum* L. fruits are being used to cure a variety of ailments, including irritation, gastritis, coughing, bronchitis, nausea, diarrhea, constipation, gout, rheumatism, sporadic fevers, as well as giddiness (Sahib et al. 2013). It's also been known to be an effective carminative, antispasmodic, and expectorant, or that's used as a treatment for arthritis or rheumatism. Antiemetic, antibacterial, emmenagogue, as well as nerve relaxing are some of the other applications (Duke et al. 2002).

Table 3. Traditional uses of different parts of *C. sativum* L.

Parts used	Traditional uses in folk medicine	Reference
Leaves and Fruits	The juice of the leaves is being used to cure skin problems as well as stress or sleeplessness. Seed decoction is being used as an eyewash to treat eye disorders. Ground seed paste is applied to the forehead to relieve migraines, as well as an infusion of seeds, has been used as a gargle for oral infections as well as gum disease. The fruits are being used as carminatives, stimulants, diuretics, tonics, stomachics, aphrodisiacs, as well as to cure convulsions, sleeplessness, and anxiousness, or to enhance appetite.	Sahib et al. (2013)
	The ethnic group uses both components of the plant as digestion remedies.	Pieroni and Gray (2008)
Fruits	Fruits have been used to cure bladder infections as well as prostate disorders, and to aid recovery or enhance feed intake after childbirth.	Abu Rabia (2005)
	Sore throat, bronchitis, nausea, gastritis, constipation, migraine, diarrhea, arthritis, spam, gout, fevers, as well as giddiness are all treated with coriander seeds in Medicinal herbs. They're also utilized as anti-inflammatories, gastrointestinal enhancers, as well as to cure gut problems.	Chaudhry and Tariq (2006)
	The seeds' infusions are utilized as anti-diabetic medicines, while extracts are utilized as anti-fertility as well as contraceptive agents.	Al Rowais (2002)

The seeds are used to cure indigestion in Ayurvedic medicine with caraway as well as cardamom seeds, or in eastern medicine with fennel, or anise seeds (Aggarwal and Kunnumakkara 2009). The herb is used to cure bloating, dysentery, constipation, cough, gastrointestinal issues, jaundice, and nausea particularly in the northern areas (Gilgit) (Khan and Khatoon 2008).

It's also used as an analgesic, a coolant, a stimulant, as well as a laxative all across the globe (Chaudhry and Tariq 2006).

C. sativum L. seeds are used to cure digestive problems, anorexia, gastrointestinal cramps, influenza with little sweating, bad breath, and odors from the genital region in traditional Medicine (Leung and Foster 1996), whereas coriander seeds are commonly consumed in Germany as medicinal teas and as ingredients in carminatives as well as laxatives. Dyspeptic symptoms, loss of appetite, stomach irritations, and bowel distress are the most common uses. The fruit is included in European Pharmacopoeias which is used as a food additive, pest inhibitor, and arthritis treatment.

C. sativum L. has a wide range of medicinal usage in Iran, where it's being used to treat convulsions, nervousness, sleeplessness, and digestive problems. In Saudi Arabia, a small percentage of the population regularly utilizes coriander seed extract as an anti-diabetic drug, as well as the seed extract is being used to reduce fertility (Al Rowais 2002). In Asia, the essential oil from the coriander herb has traditionally been used to increase gastrointestinal processes and heal ulcers or oral problems (Sahib et al. 2013). The therapeutic uses of various aspects of *C. sativum* L. in traditional medicines are summarised in Table 3.

7.2. Functional Aspects

It is well known for its beneficial effects and also health or medicinal properties, as well as additional benefits like serving as an antibacterial agent. The antibacterial activity of coriander oil seemed unaffected by the kind of food and the heat, showing that it can function as an effective antimicrobial component towards Campylobacter jejuni in food (Rattanachaikunsopon and Phumkhachorn 2010). Antioxidant activity is the most essential and well-studied pharmacological feature. The various functional characteristics of *C. sativum* L. are shown in Table 4.

Because of its high antioxidant activity, coriander is a significant source of polyphenols as well as phytochemicals. Over accumulation of free radicals and, as a result, tissue or biomolecule degradation could be caused by reactive species (Barros et al. 2012). Coriander leaves and seeds are both rich in antioxidants, although leaves have more antioxidants than seeds (Wangensteen et al. 2004). Its high pigment concentration, mainly carotenoids, is credited with its antioxidant properties. Its extract's carotenoids have been discovered to have a stronger hydroxyl radical scavenging capacity, defending cells from oxidative damage (Peethambaran et al. 2012).

Phenolic molecules are among the most important and interesting categories of bioactive compounds. Based on the number of phenol rings and the mechanical components that bind these rings, phenolic groups can be divided into four categories. Flavonoids (anthocyanins, flavones, and isoflavones), tannins, stilbenes, and lignans are among these groups (Balasundram et al. 2006). Essential oils have recently been recognized as antioxidant compounds. Coriander essential oils have antioxidant properties. Camphor (44.99 percent), cyclohexanol acetate (cis-2- tert. butyl-) (14.45 percent), limonene (7.17 percent), and -pinene are the main constituents of its essential oil (6.37 percent). This essential oil is highly effective in invading primary and secondary oxidation products at concentrations of 0.05, 0.10, and 0.15. It was asserted that at a concentration of 0.02 percent, its impacts were nearly identical to those of BHA (butylated hydroxyanisole) (Darughe et al. 2012). Oxidation causes adverse flavor alterations as well as nutritional (vitamin) depletion, resulting in inconsistency as well as appearance changes in high-fat products in the food industry.

Table 4. Different functional properties of *C. sativum* L.

Functional activity	Compounds involved	References
Antioxidant activity	Phenolics	Wangensteen et al. (2004)
Metal detoxification	Hydroxyl groups	Karunasagar et al. (2005)
Oxidative damage protection	Flavonoids (Anthocyanins, tannins, lignins)	Balasundarum et al. (2006)
Inhibition of oxidative products	Terpenoids	Bhanger et al. (2008)
Antimicrobial activity	Coriander oil	Rattanachaikunsopon and Phumkhachorn (2010)
Antiglycemic activity	Aqueous seed extracts	Deepa and Anuradha (2011)
Antianxiety effect	Coriander extract	Mahendra (2011)
Inhibition of oxidative products	BHA	Darughe et al. (2012)

Lipid peroxidation causes over-accumulation of free radicals, resulting in the development of rancidity, unpleasant taste, and odors as well as changes in color and losses related to nutritional value. The use of antioxidants reduces oxidative rancidity (Bhanger et al. 2007). Darughe et al. (2012) studied the antioxidant effects of coriander essential oils in the cake. It was found that the antioxidant effect of coriander essential oils may be due to the presence of terpenoid components (camphor, limonene, α-pinene, and geraniol). This essential oil due to its radical scavenging activity can be used as a natural antioxidant to enhance the shelf stability of many foods (Ramadan et al. 2003).

C. sativum L. can eliminate harmful metals from the body, therefore it can be utilized as a natural cleansing agent. *C. sativum* L. contains chemical components that bind to harmful metals and eliminate them from cells (Momin et al. 2012). This herb is highly efficient at removing inorganic (Hg^{2+}) and methyl mercury (CH_3Hg^+) from aqueous solutions, according to Karunasagar et al. (2005). This was owing to the carboxylic group's ability to bind to mercury. This study revealed that sorbent may effectively remove both inorganic and methyl mercury from polluted water. According to Kansal et al. (2011), coriander reduced the plethora generation of reactive oxygen species produced by lead nitrate.

The bactericidal activity of aliphatic (2E)-alkenals and alkanals isolated from fresh coriander leaves was discovered against *Salmonella choleraesuis* spp. *choleraesuis* ATCC 35640. With a minimum bactericidal concentration (MBC) of 6.25 g/mL (34 M), (2E)-Dodecenal has been the most efficient against this food-borne bacteria, followed by (2E)-Undecenal (C11) with an MBC of 12.5 g/mL (74 M). Other functional properties of coriander promote being used as a preventative as well as the remedial herb.

Antifertility activity leading to a huge drop in serum progesterone is one of these essential functions. It is used in the manufacture of several common domestic medications for treating bed sore throats, seasonal fevers, vomiting, diarrhea, as well as gastrointestinal diseases, and also dyspepsia, worms, gout, or muscle pain. Its excellent phytonutrients are responsible for many of its curative qualities, and it is often considered a bioactive chemical repository (Rajeshwari and Bondada 2011). It also prevents stomach mucosal membrane damage in a plethora of ways, including free radical scavenging activity and the synthesis of a protective layer (Al-Mofleha 2006). Its oil could be utilized as an antimicrobial agent as well. This oil is efficient towards gram-positive as well as gram-negative bacteria, and also pathogenic fungi. Except for *Bacillus cereus* and *Enterococcus faecalis*, coriander oil has bactericidal efficacy (Silva 2011). Hyperlipidemia increases the risk of lipid oxidation metabolites accumulating in the vasculature or bone's subendothelial spaces. Atherogenic high-fat cuisines raise serum levels of oxidized lipids, which have been shown to inhibit osteogenesis as well as induce bone loss in vitro (Pirih et al. 2012).

Conclusion

Herbs and spices have been used in diets since ancient times for flavor, increasing storability, as well as promoting healing. Coriander is a remarkable herb that may be used as a condiment and also a medicinal herb. Even though the herb may be produced throughout the year, it is analyzed to improve its processability, productivity, as well as facilitate global trade. Coriander seems to be a nutritious herb, a well-known condiment, and therefore is harmless to humans. It produces two types of herbal raw resources: fruits and leaves. Coriander has long been utilized in numerous countries as a culinary element as well as for a variety of medical purposes. The leaves and fruits have a higher yield and are abundant in nutrients such as lipids, proteins, vitamins, as well as minerals. Its antimicrobial and anticancer properties are among its medicinal benefits. Antimicrobial, antioxidant, anti-mutagenic, anxiolytic, sedatives, anti-depressant, neuro-protective, anti-diabetic, diuretic, anti-hypertensive, lead-detoxifying, and gut modulatory effects have made the herb prominent in non-producing regions. Coriander's most essential and well-known advantage is its ability to act as an antioxidant. This herb is frequently recognized as an "herb of good fortune" because of its multifunctional applications and a preventative or curative effect against a variety of chronic conditions. Furthermore, the suitable method to preserve coriander is to prepare its fruits as well as leaves.

References

Abu-Rabia, Aref. (2005). "Herbs as a food and medicine source in Palestine." *Asian Pacific Journal of Cancer Prevention, 6*(3), 404.

Aggarwal, Bharat B., and Ajaikumar B. Kunnumakkara eds. *Molecular targets and therapeutic uses of spices: modern uses for ancient medicine*. World Scientific, 2009.

Al-Mofleh, I. A., A. A. Alhaider, J. S. Mossa, M. O. Al-Sohaibani, S. Rafatullah, and S. Qureshi. (2006). "Protection of gastric mucosal damage by *Coriandrum sativum* L. pretreatment in Wistar albino rats." *Environmental Toxicology and Pharmacology 22*(1), 64-69.

Al-Rowais, Norah A. (2002). "Herbal medicine in the treatment of diabetes mellitus." *Saudi medical journal, 23*(11), 1327-1331.

Balasundram, Nagendran, Kalyana Sundram, and Samir Samman. (2006). "Phenolic compounds in plants and agri-industrial by-products: Antioxidant activity, occurrence, and potential uses." *Food chemistry, 99*(1), 191-203.

Barros, Lillian, Montserrat Dueñas, Maria Inês Dias, Maria João Sousa, Celestino Santos-Buelga, and Isabel CFR Ferreira. (2012). "Phenolic profiles of in vivo and in vitro grown *Coriandrum sativum* L." *Food Chemistry, 132*(2), 841-848.

Bhanger, Muhammad Iqbal, Shahid Iqbal, Farooq Anwar, Muhammad Imran, Mubeena Akhtar, and Muhammad Zia-ul-Haq. (2008). "Antioxidant potential of rice bran extracts and its effects on stabilisation of cookies under ambient storage." *International journal of food science & technology, 43*(5), 779-786.

Bhuiyan, M. N. I., Begum, J., Sultana, M. (2009). "Chemical composition of leaf and seed essential oil of *Coriandrum sativum* L. from Bangladesh." *Bangladesh Journal of Pharmacology, 4*(2), 150-153.

Burdock, George A., and Ioana G. Carabin. (2009). "Safety assessment of coriander (*Coriandrum sativum* L.) essential oil as a food ingredient." *Food and Chemical Toxicology, 47*(1), 22-34.

Cerkauskas, R. F. (2009). "Bacterial leaf spot of cilantro (*Coriandrum sativum*) in Ontario." *Canadian journal of plant pathology, 31*(1), 16-21.

Chaudhry, N. M., and Perween Tariq. (2006). "Bactericidal activity of black pepper, bay leaf, aniseed and coriander against oral isolates." *Pakistan journal of pharmaceutical sciences, 19*(3), 214-218.

Coşkuner, Yalçın, and Erşan Karababa. (2007). "Physical properties of coriander seeds (*Coriandrum sativum* L.)." *Journal of Food Engineering, 80*(2), 408-416.

Czygan F. C., D. Frohne C. Holtzel, A. Nagell and H. J. Pfander. (2001). Norman Grainger Bisset and max Wichtl. Herbal Drugs and Phytopharmaceuticals. A *Handbook for practices on a scientific basic with reference to German commission and Monographs*. Second International edition, medpharm scientific Publishers, 159-160.

Darughe, F., M. Barzegar, and M. A. Sahari. (2012)."Antioxidant and antifungal activity of Coriander (*Coriandrum sativum* L.) essential oil in cake." *International Food Research Journal, 19*(3), 1253-1260.

Das, N. (1998). "Reaction of coriander varieties to root-knot nematode (*Meloidogyne incognita* Chitwood)." *Annals of Agricultural Research, 19*(1), 94-95.

Deepa, B., and C. V. Anuradha. (2011). "*Antioxidant potential of Coriandrum sativum L. seed extract.*"

Dennis, J. I. (2003). "New disease of coriander in Australia associated with a *Microdochium* species." *Plant Pathology, 52*(3).

Diederichsen, Axel. (1996). *Coriander: Coriandrum sativum L.* Vol. 3. Bioversity International.

Divya, Peethambaran, Bijesh Puthusseri, and Bhagyalakshmi Neelwarne. (2012). "Carotenoid content, its stability during drying and the antioxidant activity of commercial coriander (*Coriandrum sativum* L.) varieties." *Food Research International, 45*(1), 342-350.

Duke, J. A., Bogenschutz-Godwin, M. J., Pu Celliar, J., Duke, P. A. K. (2002). *Handbook of medicinal herbs* 2nd edn, CRC Press: Boca Raton, 222.

Garibaldi, A. Gilardi, G. Gullino, M. L. (2010). "First report of collar and root rot caused by *Pythium ultimum* on coriander in Italy." *Plant Disease, 94*:1167.

Hwang, E., Lee, D. G., Park, S. H., Oh, M. S., & Kim, S. Y. (2014). "Coriander leaf extract exerts antioxidant activity and protects against UVB-induced photoaging of skin by regulation of procollagen type I and MMP-1 expression." *Journal of medicinal food, 17*(9), 985-995.

Kansal, Leena, Veena Sharma, Arti Sharma, Shweta Lodi, and S. H. Sharma. (2011). "*Protective Role of Coriandrum sativum Extract against lead nitrate induced oxidative stress and tissue damage in the liver and kidney in male mice. a.*"

Karunasagar, D., MV Balarama Krishna, S. V. Rao, and J. Arunachalam. (2005). "Removal and preconcentration of inorganic and methyl mercury from aqueous media using a sorbent prepared from the plant *Coriandrum sativum*." *Journal of hazardous materials, 118*(1-3), 133-139.

Keshwal, R. L., and R. K. Khatri. (1998). "Reaction of some high yielding varieties of coriander to powdery mildew." *Journal of Mycology and Plant Pathology (India).*

Khan, I. U., W. Dubey, and V. Gupta. (2014). "Taxonomical aspects of Coriander (*Coriandrum sativum* L.)." In *International*, pp. 9926-9930.

Khan, Sher wali, and S. U. R. A. Y. Y. A. Khatoon. (2008). "Ethnobotanical studies on some useful herbs of Haramosh and Bugrote valleys in Gilgit, northern areas of Pakistan." *Pakistan Journal of Botany, 40*(1), 43.

Khare, M. N., S. P. Tiwari, and Y. K. Sharma. (2017). "Disease problems in the cultivation of coriander (*Coriandrum sativum* L.) and their management leading to production of high quality pathogen free seed." *International Journal of Seed Spices, 7*(1), 1-7.

Krishnan K. S. (2001). *Dictionary of Indian Raw material and industrial plants The Wealth of India, first supplement series* (Raw materials); Volume-2: Ci – Cy; National institute of science communication and information, Resources, council of scientific and industrial Research, (CSIR), New Delhi, India. 203-206.

Leung A. Y., Foster S. (1996). *Encyclopedia of common natural ingredients used in food, drugs and cosmetics* 2nd Ed. John Wileys & Sons: New York.

Mahendra, Poonam, and Shradha Bisht. (2011). "Anti-anxiety activity of *Coriandrum sativum* assessed using different experimental anxiety models." *Indian journal of pharmacology, 43*(5), 574.

Malhotra, S. K., and B. B. Vashishtha. (2008). "Package of practices for production of seed spices." *National Research Centre on Seed Spices, Tabiji, Ajmer, India* 98.

Mandal, S., and Mandal, M. (2015). "Coriander (*Coriandrum sativum* L.) essential oil: Chemistry and biological activity." *Asian Pacific Journal of Tropical Biomedicine, 5*(6), 421-428.

Momin, Abidhusen H., Sawapnil S. Acharya, and Amit V. Gajjar. (2012). "*Coriandrum sativum*-review of advances in phytopharmacology." *International Journal of Pharmaceutical Sciences and Research,* 3(5), 1233.

Nadeem, Muhammad, Faqir Muhammad Anjum, Muhammad Issa Khan, Saima Tehseen, Ahmed El-Ghorab, and Javed Iqbal Sultan. (2013). "Nutritional and medicinal aspects of coriander (*Coriandrum sativum* L.): A review." *British Food Journal.*

Neffati, M., and Marzouk, B. (2008). "Changes in essential oil and fatty acid composition in coriander (*Coriandrum sativum* L.) leaves under saline conditions". *Industrial Crops and Production, 28*, 137-42.

Pathak Nimish, L., B. Kasture Sanjay, M. Bhatt Nayna, and D. Rathod Jaimik. (2011). "Phytopharmacological properties of *Coriander sativum* as a potential medicinal tree: an overview." *Journal of Applied Pharmaceutical Sciences, 1*(4), 20-5.

Pieroni, Andrea, and Charlotte Gray. (2008). "Herbal and food folk medicines of the Russlanddeutschen living in Künzelsau/Taläcker, South-Western Germany." *Phytotherapy Research, 22*(7), 889-901.

Pirih, F., Lu, J., Ye, F., Bezouglaia, O., Atti, E., Ascenzi, M. G., Tetradis, S., Demer, L., Aghaloo, T., Tintut, Y. (2012). Adverse effects of hyperlipidemia on bone regeneration and strength. *Journal of Bone and Mineal Research*, *27*(2), 309-318.

Rajeshwari, Ullagaddi, and Bondada Andallu. (2011). "Medicinal benefits of coriander (*Coriandrum sativum* L)." *Spatula DD, 1*(1), 51-58.

Ramadan, Mohamed F., Lothar W. Kroh, and Jörg-T. Mörsel. (2003). "Radical scavenging activity of black cumin (*Nigella sativa* L.), coriander (*Coriandrum sativum* L.), and niger (*Guizotia abyssinica* Cass.) crude seed oils and oil fractions." *Journal of agricultural and food chemistry, 51*(24), 6961-6969.

Rattanachaikunsopon, Pongsak, and Parichat Phumkhachorn. (2010). "Potential of coriander (*Coriandrum sativum*) oil as a natural antimicrobial compound in controlling Campylobacter jejuni in raw meat." *Bioscience, biotechnology, and biochemistry, 74*(1), 31-35.

Ravi, Ramasamy, Maya Prakash, and K. Keshava Bhat. (2007). "Aroma characterization of coriander (*Coriandrum sativum* L.) oil samples." *European Food research and technology, 225*(3), 367-374.

Rodeva, Rossitza, Svetlana Dacheva, and Zornitsa Stoyanova. (2010). "Wilting and root rot of coriander caused by Macrophomina phaseolina in Bulgaria." In *The Proceeding of 45th International Symposium on Agriculture*.

Sahib, Najla Gooda, Farooq Anwar, Anwarul-Hassan Gilani, Azizah Abdul Hamid, Nazamid Saari, and Khalid M. Alkharfy. (2013). "Coriander (*Coriandrum sativum* L.): A potential source of high-value components for functional foods and nutraceuticals-A review." *Phytotherapy Research, 27*(10), 1439-1456.

Silva, Filomena, Susana Ferreira, João A. Queiroz, and Fernanda C. Domingues. (2011). "Coriander (*Coriandrum sativum* L.) essential oil: its antibacterial activity and mode of action evaluated by flow cytometry." *Journal of medical microbiology, 60*(10), 1479-1486.

Singh, A. K. (2009). "Integrated management of wilt, *Fusarium oxysporum* f. sp. *coriandrii* of coriander." *Indian Journal of Plant Protection, 37*(1-2), 132-133.

Singh, H. B., A. Singh, A. Tripathi, S. K. Rai, R. S. Katiyar, J. K. Johri, and S. P. Singh. (2003). "Evaluation of Indian coriander accessions for resistance against stem gall disease." *Genetic Resources and Crop Evolution, 50*(4), 339-343.

Tripathi, A. K., A. M. Barataria, R. K. Pandya, S. Chauhan, and S. Chauhan. (1998). "Evaluation of coriander cultivars against stem gall disease." *Flora and Fauna, Jhansi, 4*, 98.

Tripathi, A. K., A. M. Bartaria, R. K. Pandya, and M. L. Tripathi. (2002). "Effect of cultural practices on stem gall disease of coriander." *Annals of Agricultural Research, 23*, 171-173.

Usami, Toshiyuki, Satomi Morii, Chizumi Matsubara, and Yoshimiki Amemiya. (2012). "Plectosphaerella rot of lettuce, coriander, and chervil caused by *Plectosphaerella pauciseptata*." *Journal of general plant pathology, 78*(5), 368-371.

Wangensteen, Helle, Anne Berit Samuelsen, and Karl Egil Malterud. (2004). "Antioxidant activity in extracts from coriander." *Food chemistry, 88*(2), 293-297.

Yeung, Edward C., and Steve Bowra. (2011). "Embryo and endosperm development in coriander (*Coriandrum sativum*)." *Botany, 89*(4), 263-273.

About the Editors

Dr. Mamta Pujari is Assistant Professor in the department of Botany, Lovely Professional University, Phagwara, Punjab (India). She completed her Post graduate study in Botany from Hemvati Nandan Bahuguna Garhwal University and Doctorate degree from Kumaon University, Nainital. She has published more than 10 research papers and 7 book chapters in internationally published volumes such as nova, Elsevier, springer etc.

Dr. Bharat Kapoor, MHM, MBA (Hospitality Management), PhD, is Assistant Professor in the department of Hotel management and Tourism at Guru Nanak Dev University, Amritsar, Punjab, India. Dr. Bharat Kapoor has 18 years of academic and industrial experience in the field of Hospitality industry. During his industrial experience he worked with Taj Group of Hotels, The Imperial New Delhi and ITC Group of Hotels in the department of Food and Beverages. As per his academic experience is concerned, he was holding the positions of HOD-Hotel Management at Lovely Professional University Jalandhar, Director-Hospitality at CT University, Ludhiana and Dean-Doctoral Research Centre at Chitkara University, Chandigarh (India). He earned his PhD in Hotel Management from Kurukshetra University, Haryana in the year 2012. He is having more than 25 publications in reputed journals and more than 10 book chapters, one patent and two books. He has organized many National and International conferences. He is a member of various professional bodies related to the field of Hospitality and Tourism and he is also a member of Board of Studies of various reputed universities in India.

Dr. Anju Joshi has been working in Govt. Degree College, Uttarakhand since 2018. She did her Undergraduate and Postgraduate studies from Kumaon University Nainital, Uttarakhand. Her area of specialization is Ethnobotany (Taxonomy). She has published more than 10 papers and 5 book chapters in various reputed journals.

Professor Neeta Pande has been working as Professor in the department of Botany, Govt College Kumaon University Nainital for the last 22 years. She completed her Post graduate and Doctorate study in Botany from Kumaon University Nainital. She has specialization in various fields of Botany like Mycology, plant pathology, Ethnobotany, biodiversity. She has published more than 50 research Papers and 20 book chapters in internationally published volumes.

Index

A

B

C